Green

Delusions

Green

An

Environmentalist

Delusions

Critique of

Radical

Martin W. Lewis

Environmentalism

Duke University Press

Durham and London

1992

Second printing
© 1992 Duke University Press
All rights reserved
Printed in the United States of America
on acid-free paper ∞
Library of Congress Cataloging-in-Publication Data
appear on the last printed page of this book.

Contents

Acknowledgments

It is difficult to acknowledge the many persons who have contributed to this effort without implicating them in the political and philosophical positions advocated here. That said, it is equally important to recognize that *Green Delusions* could never have been written without the assistance, and concerted criticism, of numerous individuals.

My closest colleague, and most demanding critic, is Karen Wigen, assistant professor of Japanese history and geography at Duke University. Although her imprint pervades the text, it is essential to note that we parted company in chapters five and six. Other academic geographers who have helped in this endeavor include James Delehanty, Stephen Frenkel, Piper Gaubatz, Deborah Hart, Marie Price, Paul Starrs, Stan Stevens, and Karl Zimmerer. I especially thank Steve Frenkel for sharing his observations on the parallel constructions of deep ecology and environmental determinism. I also offer my gratitude to my mentors in the Berkeley department of geography, James Parsons, Bernard Nietschmann, and Michael Watts, although I imagine that each will find portions of this work highly objectionable. My students at George Washington University also deserve a note of appreciation; special thanks to Gregg Bernstein, who gave invaluable help as a research assistant.

In many ways it is my friends, acquaintances, and relatives outside of academia who have given me the perspective necessary to write this book. Kathryn Lodato shared her penetrating insights into environmental philosophy and action, and Donald Boroughs was generous with his knowledge and ideas about the current state of the American economy. Joel Margolis and Valerie Russell showed me the importance of telecommunications, and gave some valuable lessons on the business world. Lyndon Peter insisted that I not ignore the current plight of American farmers when discussing the future of agriculture, and Gordon Kato provided one invaluable citation in addition to giving some helpful suggestions for the manuscript in general. I must also express my appreciation to Kim Bruno, Wendy Clark, and especially James Hynes for the best political discussions I have had in some time. Wayne and Nell Lewis, in whose house this project had its genesis, provided essential material and intellectual support. I give special thanks to Wayne Lewis for showing me, from an early age, that one can be a tough-minded liberal.

Finally, sincere thanks to Larry Malley and everyone else at Duke University Press. My association with the press has been a delightful experience from the beginning.

Green

Delusions

Introduction

The global environment currently faces two profound ideological threats. The first, and most serious, stems from the work of anti-environmentalists who would have us believe that the ecological crisis is a mirage. Such modern-day Pollyannas as Ben Wattenberg (1987), Dixy Lee Ray (1990), and Julian Simon and Herman Kahn (1984) present a comforting vision to those who shudder at the thought of the sacrifices that will be necessary to ensure the ecological health of the planet. But as their writings reveal, this cadre evidently considers human beings the only organisms worthy of consideration. To anyone who deems the survival of other species important, such "optimism" offers cold comfort indeed.

Despite ritual protestations to the contrary, these attacks on ecological concern are blatant enough to receive fierce opposition from all environmental quarters. But there is a much less visible ideological threat at work as well, one that masquerades under the mantle of environmentalism itself. In my view, many of the most committed and strident "greens" unwittingly espouse an ill-conceived doctrine that has devastating implications for the global ecosystem: so-called radical environmentalism. It is the purpose of this book to distinguish the five main variants of eco-extremism currently being forwarded, to expose the fallacies upon which such views ultimately flounder, and to demonstrate that the policies advocated by their proponents would, if enacted, result in unequivocal ecological catastrophe.

I would emphasize at the onset that in attacking eco-radicalism I seek not to assail but rather to defend the broad-based environmental movement. The great majority of environmental organizations shun extremism, remaining committed to reforming our economy and society through dogged work within normal political and legal channels. Such groups deserve, and require, widespread public praise and support. But eco-extremism, based on a doctrine of radical ecological salvation, has been gaining strength rapidly over the past several years. In fact, the environmental mainstream is itself now under assault from the eco-radicals, as extremists denounce such steadfast organizations as the Sierra Club or the Audubon Society for having "sold out to the de-spoilers." It is precisely to defend environmentalism—a term that many radicals tellingly disdain as reeking of accommodation—that I have chosen to speak out against environmental extremism. The latter must be countered, for in seeking to dismantle modern civilization it has the potential to destroy the very foundations on which a new and ecologically sane economic order must be built.

Throughout this work I will argue vigorously, at times fiercely, against the many doctrines of green radicalism. I do this because I am convinced that such ideas are beginning to lead the environmental movement toward self-defeating political strategies, preventing society from making the reforms it so desperately needs. This does not by any means entail denouncing all of the committed activists holding eco-radical views. To the contrary, many of them deserve praise for their genuine concern and for the personal sacrifices they are making on behalf of the earth. It is rather their ill-conceived *ideas* with which I am concerned. If at times my aspersions are caustic, it is because I have had to battle against these seductive ideas myself. Until a few years ago, I too endorsed all of the main platforms of the radical greens.

■ Environmentalism: Radical and Mainstream Positions

Radical environmentalism is a multistranded philosophy, the several variants of which will be explored in detail in chapter two. But beneath their many differences, eco-radicals concur in one central proposition: that human society, as it is now constituted, is utterly unsustainable and must be reconstructed according to an entirely different socioeconomic logic. The dominant school of radical green thought argues specifically that a sustainable society must be small in scale and modest in technology. Eco-radicals, therefore denounce anyone seeking merely to reform, and thus perpetuate, a society that they regard as intrinsically destructive if not actually evil.

The Postulates of Eco-Radicalism

Radicalism, as the term is employed here, is the notion that society must be attacked "at the root" and reinvented, whether through revolution, enlightenment (the massive ideological conversion of the populace), or the rebuilding that would follow upon an ecological holocaust. Most radical greens, to their credit, forswear revolution and hope to avoid Armageddon. But the movement as a whole consistently questions and often disparages the struggle for environmental reform within the framework of liberal democracy (Manes 1990:19; Pell 1990). Such reform, extremists argue, would only allow a terminally sick society to persist in its fundamentally destructive habits, forestalling the drastic rebuilding necessary for long-term survival.

The dominant version of radical environmentalism rests on four essential postulates: that "primal" (or "primitive") peoples exemplify how we can live in harmony with nature (and with each other); that thoroughgoing decentralization, leading to local autarky, is necessary for ecological and social health; that technological advance, if not scientific progress itself, is inherently harmful and dehumanizing; and that the capitalist market system is inescapably destructive and wasteful. These views, in turn, derive support from an underlying belief that economic growth is by definition unsustainable, based on a denial of the resource limitations of a finite globe.

In accordance with these tenets, radical environmentalists would have us abandon urban, industrial, capitalist civilization and return to the earth. Here our modern social maladies will be healed as we find true harmony between land and life. The call is for a simpler existence and a more direct relationship between humanity and nature, one in which natural landscapes are transformed by human agency as little as possible. People must never be so arrogant as to "manage" nature; rather, nature should always be allowed to exemplify its own essential harmonies.

Within the human community, according to this doctrine, economic relationships likewise should be unmediated. Radical green thinkers usually consider trade to be alien and finance positively wicked. Third World countries are urged not only to shun industrialization but also to isolate themselves from the intrinsically exploitative global economic system. According to one prominent theory, only village-based, low-tech, agrarian development strategies will allow poor societies to fulfill basic human needs.

Most importantly, eco-radicals inform us that economic growth must simply come to an end. The larger the human economy becomes, the more nature suffers. Whereas moderate environmentalists see some benefits to economic growth (a more prosperous society being able to afford more environmental protection), green stalwarts consider this proposition self-canceling. "We do not cure treatment-induced diseases by increasing the treatment dosage," writes Herman Daly (1977:101). Or, as Arne Naess (1989:211), the founder of the "deep ecology" movement, frames it: "One per cent increase in Gross National Product today inflicts far greater destruction of nature than one per cent 10 or 20 years ago because it is one per cent of a far larger product. And the old rough equivalency of GNP with 'Gross National Pollution' still holds. And the efforts to increase GNP create more formidable pressures against ecological policies every year." Fortunately, we need not fear that the end of economic growth will diminish human well-being—quite the contrary, for our obsession with economic growth, and more generally, the possession of material goods, has only hidden from our view an underlying spiritual impoverishment (Daly 1977:44). By relinquishing our mad drive for progress we may at last begin to build a truly humane and genuinely prosperous community.

Contemporary Environmentalism: Moderate and Radical
While these radical notions seem to be beginning to dominate the environmental discourse, they by no means monopolize it. In fact, a pragmatic philosophy continues to inform most of the movement's large organizations. In *An Environmental Agenda for the Future*, for example, leaders of the country's ten most influential conservation groups explicitly accept capitalism and expressly endorse economic growth (Cahn 1985). Robert Repetto and others at the World Resources Institute argue powerfully that environmental health depends on sustainable development—on ecologically sound economic programs that aim to increase human "wealth and well-being" (Repetto 1986:10). Other advocates of sustainable development, such as R. Kerry Turner, stress "the need to view environmental protection and continued economic growth . . . as mutually compatible and not necessarily conflicting objectives" (Turner 1988:5). Piasecki and Asmus (1990) similarly propose a moderate course in seeking "environmental excellence," while Henning and Mangun (1989:203) claim that the presumed conflict between economic growth and environmental salvation rests on a false dichotomy. Still other theorists might be described as "techno-environmentalists." Oppenheimer and Boyle (1990), for example, propose that technical advances in power generation and in transportation systems offer the best hope for protecting the atmosphere.

But, as Peter Borrelli (1988:19) shows, the larger environmental community increasingly suspects that the mainstream organizations have compromised too much, falling into the trap of "creeping conservatism." The moderate stance of the established lobbying organizations seems far too weak to those who are most distraught over the planet's failing health. Seeking to avoid this pit, the most vociferous of the mainstream groups, Greenpeace, has recently called for a grassroots revolution against pragmatism and compromise (*Greenpeace*, July–August 1990). Many radical greens have even begun to view institutional conservation groups as enemies, or at best as allies of convenience only. As one noted radical put it, "The worldview of the executive director of the Sierra Club is closer to that of James Watt or Ronald Reagan [than it is to our own]" (quoted in Manes 1990:225). Such views appear to have seized the movement's heart already, and they may soon be poised to grasp its political initiative as well.

Radicalism's Fillip: The Environmental Crisis

The attractions of a radical perspective are very real. As signs of impending disaster mount, it is not at all surprising that concerned individuals should seek forceful action—and the strident philosophies that would justify it. Those who espouse radical notions, though gravely mistaken in the solutions they propose, are not wrong about the magnitude of the problem.

The signals of impending disaster are many. Global warming and especially the depletion of stratospheric ozone are extraordinary threats to life on earth. The need to graduate from fossil fuels and move into a solar age is paramount, as is the necessity to recycle, rather than discard, our material artifacts. Equally worrisome are contemporary agricultural practices. The bombarding of croplands with life-destroying chemicals is unconscionable and, in the long run, economically self-defeating as well.

Perhaps most appalling is the mass extermination of plant and animal species presently occurring in many reaches of the world. Environmental extremists do clearly occupy the moral high ground here in arguing that life should be protected for its own sake, rather than merely for the benefits that it might someday proffer to humanity. Furthermore, it is not just extinction per se that should concern us; as Soulé (1980) has shown, the reduction of the numbers and the geographical ranges of large mammals has effectively ended their ongoing evolution. Wildlife must be not merely saved but restored—a monumental task, to say the least.

Environmental radicals also have a compelling point when they argue that the earth's human population is already too large. While the Malthusian vision of humanity soon exceeding global carrying capacity is probably fantasy, the simple fact that some 30 percent of the planet's primary production is now devoted to sustaining our species (Ehrlich 1988:23) is reprehensible. I too believe that we have a moral obligation to share this world with its other inhabitants, which the very weight of our own numbers is making increasingly difficult.

And finally, all environmentalists should applaud certain nonviolent acts of civil disobedience undertaken by eco-extremists. Whether they are harassing Icelandic whaling vessels or chaining themselves to ancient trees in the Pacific Northwest, green radicals show great courage in trying to stop needless destruction. But in lauding these actions one is by no means obligated to endorse the beliefs that presently lie behind them. In fact, as I will argue, only by renouncing those beliefs can we begin to create a truly effective environmental movement.

■ The Fatal Flaws of Eco-Radicalism

Those who condemn radical environmentalism usually do so on the grounds that it represents a massive threat to human society, to civilization, and material progress as we have come to know them. This is hardly surprising, since many eco-radicals proudly claim this to be their intent. But the core argument of this work is that green extremism should be more deeply challenged as a threat to nature itself. Because this thesis is seemingly paradoxical, it is necessary to outline in unambiguous terms how nature's most fervent champions could also unwittingly be among its most dangerous foes.

A Green Threat to Nature
The most direct way in which eco-extremists threaten the environment is simply by fueling the anti-environmental countermovement. When green radicals like Christopher Manes (1990) call for the total destruction of civilization, many begin to listen to the voices of reaction. Indeed, the mere linking of environmental initiatives to radical groups such as Earth First! often severely dampens what would otherwise be widespread public support (see Gabriel 1990:64).

As radicalism deepens within the environmental movement, the oppositional anti-ecological forces accordingly gain strength. The Center for the Defense of Free Enterprise, a think tank for the so-called wise use movement, has, for example, recently published a manifesto calling for such outrages as the opening of all national parks to mineral production, the logging of all old-growth forests, and the gutting of the endangered species act. This group's ideologues contend that certain environmental philosophies represent nothing less than mental illnesses, a theory anonymously propounded in the "intellectual ammunition department" of their *Wise Use Memo* (Center for the Defense of Free Enterprise 1990:2). Even more worrisome is the fact that a former high-ranking CIA agent is now spreading rumors that environmental scientists are presently attempting to concoct a virus that could destroy humankind (See "Tale of a Plot to Rid Earth of Humankind," *San Francisco Examiner*, April 14, 1991: A-2). My fear is that if green extremism captures the environmental movement's upper hand, the public would be much less likely to recognize such a claim as paranoid fantasy; while a handful of eco-radicals would be happy to destroy humanity, such individuals also reject science and thus would never be able to act on such convictions.

More frightening, and more immediate, is the specter of a few radicals actually opposing necessary environmental reforms. Such individuals conclude that "reform environmentalism" is "worse than useless because by correcting short-term symptoms it postpone[s] the necessary reconstruction of the entire human relationship with the natural world" (Nash 1989:150). From here it is a short step to argue that reform would only forestall an ecological apocalypse—which some evidently believe is a necessary precondition for the construction of an environmentally benign social order. The insanity of pushing the planet even closer to destruction in order to save it in the future should be readily apparent.

While such are the fantasies only of the most moonstruck extremists, even moderate radicals (if one may be permitted the oxymoron) espouse an ideology that would preclude the development of an ecologically sustainable economy. Most environmentalists, for instance, aver that a sustainable economy must be based on solar power. Yet the radicals' agenda, calling for total decentralization, deurbanization, economic autarky, a ban on most forms of high technology, and the complete dismantling of capitalism, would not only prevent future improvements in solar power but would actually destroy the gains that have already been made. While most radical greens embrace "appropriate technologies" (just as anti-environmentalists denounce "pollution"), their program would, if enacted, undercut the foundations of all technological research and development. Appropriate technology, in fact, often turns out to mean little more than well-engineered medieval apparatuses: we may expect crude mechanical power from the wind, but certainly not electricity from the sun. Equally important, the systematic dismantling of large economic organizations in favor of small ones would likely result in a substantial increase in pollution, since few small-scale firms are able to devise, or afford, adequate pollution abatement equipment.

Radical environmentalism is similarly culpable on the issue of demography. While adherents argue powerfully that the reduction of human numbers is necessary for ecological health, all indications are that their envisaged social program would only lead to further growth. Reverting to small, self-sustaining rural communities would only recreate the conditions that have historically led, in most cases, to steady population gains. Precluding industrialization in the Third World is also likely to ensure that high fertility rates remain the global norm.

Radical environmentalism presents yet another threat to the earth in its unyielding opposition to the human management of nature. The notion that "nature knows best" is meaningless in a world already re-made to anthropogenic contours. Pristine nature is nonexistent—and has been, except perhaps in a few remote islands, for thousands of years (Goudie 1981; Simmons 1989). Only thoroughgoing human intervention can preserve remaining biological diversity until the time comes when a more advanced human society can begin to let nature reclaim more of the earth (see, for example, Janzen 1988). In particular, while some eco-radicals evidently believe that "liberating zoos" will help "save the earth" (G. Smith 1990), responsible environmentalists fully realize that such actions would, at present, only ensure the extinction of numerous endangered animal species currently being bred in zoological gardens (Conway 1988).

Finally, the radical green movement threatens nature by advocating a return to the land, seeking to immerse the human community even more fully within the intricate webs of the natural world. Given the present human population, this is hardly possible, and even if it were to occur it would result only in accelerated destruction. Ecological philosophers may argue that we could follow the paths of the primal peoples who live in intrinsic harmony with nature, but they are mistaken. Tribal groups usually do live lightly on the earth, but often only because their population densities are low. To return to preindustrial "harmony" would necessarily entail much more than merely decimating the human population.

Yet unless our numbers could be reduced to a small fraction of present levels, any return to nature would be an environmental catastrophe. The more the human presence is placed directly on the land and the more immediately it is provisioned from nature, the fewer resources will be available for nonhuman species. If all Americans were to flee from metropolitan areas, rural populations would soar and wildlife habitat would necessarily diminish.

An instructive example of the deadly implications of returning to nature may be found when one considers the issue of fuel. Although more common in the 1970s than the 1990s, "split wood not atoms" is still one of the green radicals' favored credos. To hold such a view one must remain oblivious to the clearly devastating consequences of wood burning, including suffocating winter air pollution in the enclosed basins of the American West, widespread indoor carbon monoxide poisoning, and the ongoing destruction of the oak woodlands and savannahs of California. If we were all to split wood, the United States would be a deforested, soot-choked wasteland within a few decades. To be sure, the pollution threat of wood stoves can be mitigated by the use of catalytic converters, but note that these are technologically sophisticated devices developed by capitalist firms.

If the most extreme version of the radical green agenda were to be fully enacted without a truly *massive* human die-off first, forests would be stripped clean of wood and all large animals would be hunted to extinction by hordes of neo-primitives desperate for food and warmth. If, on the other hand, eco-extremists were to succeed only in paralyzing the economy's capacity for further research, development, and expansion, our future could turn out to be reminiscent of the environmental nightmare of Poland in the 1980s, with a stagnant economy continuing to rely on outmoded, pollution-belching industries. A throttled steady-state economy would simply lack the resources necessary to create an environmentally benign technological base for a populace that shows every sign of continuing to demand electricity, hot water, and other conveniences. Eastern Europe shows well the environmental devastation that occurs when economic growth stalls out in an already industrialized society.

The Shaky Foundations of Radical Environmentalism
In the pages that follow I will argue that each of the four essential postulates of radical environmentalism outlined above is directly contradicted by the empirical record. "Primal" economies have rarely been as harmonized with nature as they are depicted; many have actually been highly destructive. Similarly, decentralized, small-scale political structures can be just as violent and ecologically wasteful as large-scale, centralized ones. Small is sometimes ugly, and big is occasionally beautiful. Technological advance, for its part, is clearly necessary if we are to develop less harmful ways of life and if we are to progress as a human community. And finally, capitalism, despite its social flaws, presents the only economic system resilient and efficient enough to see the development of a more benign human presence on the earth.

But a critique of these notions, however sound, misses the fundamental point. Ultimately, green extremism is rooted in a single, powerful conviction: that continued economic growth is absolutely impossible, given the limits of a finite planet. Only if this notion is discredited can the edifice of eco-radical philosophy be shaken.

It can logically be shown that the supposed necessity of devising a steady-state economy is severely misconstrued. Economic growth, strictly speaking, is defined as an increase in the *value* of goods *and* services produced. Yet as noted almost twenty years ago by Mancur Olson (1973:4), radical greens have a significantly different conception, one that largely ignores services and that substitutes mass for value. To read some of their tracts, one could only conclude that economic growth requires producing ever larger quantities of steel, wheat, and similar material goods (for example, Ornstein and Ehrlich 1989:227).

The more sophisticated advocates of a steady-state economy do, however, take care to distinguish between qualitative economic growth (which they sometimes support) and quantitative economic growth (which they denounce), holding ultimately that only human bodies and material artifacts must be held at a constant level (Daly 1977). But in the end, the bias against even nonconsumptive forms of economic growth remains. The provision of services must stabilize, warns Daly (1977:17–18), because services too require physical maintenance, while Kassiola (1990:121) cautions that services are further suspect because they are "part of the materialist value structure of industrialism." Porritt (1985: 37, 183) even worries about the economic expansion entailed by the growth of the pollution abatement and recycling industries.

Services do indeed require maintenance. Yet it is appears that as services become more sophisticated, they often require less physical support rather than more. More importantly, recent economic progress has come to demand a certain dematerialization of value, based on miniaturization and the development of lightweight, energy efficient, composite materials. Growth has also been repeatedly stimulated as relatively abundant resources are substituted for rare ones, the replacement of copper wire by fiber optic (glass) cables being a prime example. While the global economy certainly cannot grow indefinitely in *volume* by pouring out an ever mounting cavalcade of consumer disposables, it *can* continue to expand in *value* by producing better goods and services ever more efficiently.

As I shall argue repeatedly throughout this work, economic growth of this type is absolutely essential. Only a strongly expanding economic base can generate the capital necessary to retool our economy into one that does not consume the earth in feeding itself. Ecological sanity will be expensive, and if we cannot pay the price we may well perish. This proposition is even more vital in regard to the Third World; only steady economic expansion can break the linkages so often found in poor nations between rural desperation and land degradation. Genuine development, in turn, requires both certain forms of industrialization as well as participation in the global economy.

The major portion of this work is devoted to examining and criticizing the key postulates of radical environmentalism. To put it bluntly, I will argue that the foundations of green extremism are constructed upon erroneous ideas fabricated from questionable scholarship. Radical environmentalism's ecology is outdated and distorted, its anthropology stems from naive enthusiasms of the late 1960s and early 1970s, and its geography reflects ideas that were discredited sixty years ago. Moreover, most eco-radicals show an unfortunate ignorance of history and a willful dismissal of economics.

The Strategic Blundering of Eco-Radicalism

Such are the bases of the ideological critique. At the same time, it is essential to examine the realm of radical environmental action. Here I will argue that eco-radical political strategy, if one may call it that, is consummately self-defeating. The theoretical and empirical rejection of green radicalism is thus bolstered by a series of purely pragmatic objections.

Many eco-radicals hope that a massive ideological campaign can transform popular perceptions, leading both to a fundamental change in lifestyles and to large-scale social reconstruction. Such a view is highly credulous. The notion that continued intellectual hectoring will eventually result in a mass conversion to environmental monasticism (Roszak 1979:289)—marked by vows of poverty and nonprocreation—is difficult to accept. While radical views have come to dominate many environmental circles, their effect on the populace at large has been minimal. Despite the greening of European politics that recently gave stalwarts considerable hope, the more recent green plunge suggests that even the European electorate lacks commitment to environmental radicalism. In the United States several decades of preaching the same eco-radical gospel have had little appreciable effect; the public remains, as before, wedded to consumer culture and creature comforts.

The stubborn hope that nonetheless continues to inform green extremism stems from a pervasive philosophical error in radical environmentalism. As David Pepper (1989) shows, most eco-radical thought is mired in idealism: in this case the belief that the roots of the ecological crisis lie ultimately in *ideas* about nature and humanity. As Dobson

(1990:37) puts it: "Central to the theoretical canon of Green politics is the belief that our social, political, and economic problems are substantially caused by our intellectual relationship with the world" (see also Milbrath 1989:338). If only such ideas would change, many aver, all would be well. Such a belief has inspired the writing of eloquent jeremiads; it is less conducive to designing concrete strategies for effective social and economic change.

It is certainly not my belief that ideas are insignificant or that attempting to change others' opinions is a futile endeavor. If that were true I would hardly feel compelled to write a polemic work of this kind. But I am also convinced that changing ideas alone is insufficient. Widespread ideological conversion, even if it were to occur, would hardly be adequate for genuine social transformation. Specific policies must still be formulated, and specific political plans must be devised if those policies are ever to be realized.

Many of the more sophisticated eco-radicals would agree with this notion. But even the political moves advocated by the more savvy among them remain committed to a radicalism that the great majority of the American public finds unpalatable. Radical green strategists may call for alliances with new social movements or with radical political parties, but even a concerted coalition of the disaffected would be unable to approach the critical mass needed to gain effective power. And several radical thinkers have proposed that much narrower constituencies form potentially eco-revolutionary groups that might lead society as a whole to its necessary transformation. According to one theory, only the unemployed can seek real change, rather than just a redistribution of spoils, because only they do not participate in the wicked system (Dobson 1990:163). Although this represents a fringe view, the general process of seeking ever more radical foundations for social reinvention leads eco-extremists to reduce their own potential bases for political power to ever more minuscule, and powerless, groups. At the same time, most green extremists overtly denounce more moderate environmentalists who are willing to seek compromises with individuals or groups of opposing political philosophies. Since compromise, in one form or another, is necessary for any kind of effective political action, the quest for purity will in the end only undercut the prospects for change.

Even moderate environmentalists often adopt an unnecessarily exclusive political strategy. Robert Paehlke, whose *Environmentalism and the Future of Progressive Politics* stands as a monument to reason within the field, insists on attaching the movement firmly to the traditional left, urging environmentalists to appeal primarily to "industrial workers, public servants, and those employed in health, education, and the arts" (1989:276, 263). Since in the United States this traditional liberal constituency by itself has no immediate chance of gaining national power, such a tactic would again only diminish the prospects for much needed reform. At the same time, the pernicious fear of compromise seriously diminishes the possibility of creating a broader coalition for environmental action. Barry Commoner, for example, warns environmentalists that if they compromise with corporations they may become "hostages" and eventually even assume "the ideology of [their] captors" (1990:177). The end result of this kind of thinking—to which we are painfully close in the United States—is an ideological stalemate in which opposed camps are increasingly unable even to communicate. In such a political environment, the creation of an ecologically sustainable society becomes little more than an impossible dream.

Against Naive Globalism

Finally, while radical environmentalists are right to argue that the most pressing ecological problems are global in scale, the need for international cooperation does not mean that one can ignore the existing geopolitical framework of competing sovereign states. Leading countries—those boasting economic and technological prowess—will unavoidably play key roles in determining whether an ecologically sustainable socioeconomic order is ever to be created. By merely condemning rather than coming to terms with international competition, radical greens again undermine the prospects for creating a sustainable global economy.

While the United States has for some time been the world's predominant economic power, the 1980s saw that dominance severely challenged, especially by Japan. In attempting to discern the possible future economic-environmental directions to be taken by global society, it is therefore necessary to examine the relative positions and current trajectories of both countries.

The United States is currently hesitating on the threshold of a high-tech future. Its economy is in perilous straits, a vocal minority of its population wishes to dismantle the engines of technological growth, and its financial structures remain deeply biased against long-range planning and growth. Japan, on the other hand, is rushing headlong, yet with some foresight, into an ever more mechanized and synthesized future. The question is no longer whether America can remain the world's foremost economic power, but rather whether it can hope to be Japan's equal. Japan, with half our population, now invests more money annually in its basic industrial infrastructure than does the United States, and America's premier research laboratories are beginning to do as much work for Japanese as for U.S. firms. Political developments within this country may well turn out to be irrelevant to the global economic system. As Paul Krugman (1990:41) bitterly writes, "The vision of the United States as a giant Argentina may be unlikely, but no one should dismiss it out of hand."

By attacking the foundations of scientific research and technical development, radical environmentalism would hasten the decline of the United States relative to Japan. In so doing, I believe it would threaten the ecological future of the planet. Rough parity between the United States and Japan is highly desirable, in part because the two countries' environmental strengths and weaknesses are mirror images of each other. While Japan is far ahead in basic energy and resource efficiency, Americans as a whole are more concerned about the preservation of nature—for nature's sake—than are the Japanese. Japanese protests over such outrages as the mass slaughtering of dolphins are the merest of whispers. In a world economy dominated by Japan, energy and resource conservation would likely be high priorities, but not the preservation of biological diversity.

The best hope, economically as well as environmentally, lies in increasing cooperation between the United States and Japan as equals (see Rosecrance 1990). But unless the American economy is restructured to enhance long-range investments—and unless the American public unreservedly embraces continuing advances in technology—the best position we will be able to maintain will be that of a resource-rich but technologically dependent junior partner. Even if eco-extremists were to gain power in the United States, under such circumstances their ability to influence the evolutionary path of global society would be nil.

Environmentalism's Current Challenge

At present, radical environmentalism is a marginal movement that presents little threat to the status quo. Although the radicalized intellectuals of the Vietnam generation have been almost universally converted to one or another variety of eco-extremism, their influence over the electorate at large remains minimal.

One hears repeated laments that the generation coming of age now is more concerned with high wages and career security than with social change. Yet it is precisely to this cynical generation that an effective environmental philosophy must appeal. To galvanize a young electorate whose primary political act to date has been to throw in its lot with capitalism, environmentalism must reconstitute itself with an entirely new philosophical foundation.

■ Toward a Promethean Environmentalism

The radical environmentalist view sketched above is unabashedly Arcadian, calling for a return to a simpler, rural mode of life. Although this pastoral archetype has long inspired romantic visions of an alternative society, few today are aware of the fate of the historical Arcadia. While poets praised the region's rusticity, the unsophisticated Arcadians were easily bullied by Spartan aggressors. Finally, in 370 B.C.E. the Theban general Epaminondas united the southern Arcadian villagers into a new polis in order to form a bulwark against further Spartan expansion. Ironically, it was this Arcadian city that first bore the name of Megalopolis.

Despite Arcadia's sorrows, the dream associated with its name has never died. Especially since Rousseau, disaffected intellectuals have looked longingly to Arcadia as a symbol of the countryside left behind. Unfortunately, not only is Arcadia impossible to reclaim, but it never really existed as imagined in the first place. The Arcadian myth is based on a sanitized picture of nature, one from which labor and suffering have been conveniently removed (Pepper 1989:85). If we are to establish a realistic environmental movement we must begin to think in terms of a different classical archetype.

This work accordingly advocates what I call a Promethean environmentalism, one that embraces the wildly creative, if at times wildly destructive, course of human ascent. Our future lies not in abandoning technology, but in harmonizing it to a new environmental vision. We should, and we will, continue to burn Prometheus's flame—but we must learn to do so as responsible adults rather than as pyromaniacal adolescents.

There is nothing new in a Promethean approach to environmental problems; indeed, this is the hallmark of the technocratic stance of writers like Herman Kahn and Julian Simon—individuals I have already classified as anti-environmentalists. But to the technologically oriented anti-environmentalist, progress is valued strictly for human purposes, while nature is seen as little more than raw stock for the human economy. Such writers thus have few environmental concerns beyond their hope that new technologies may allow us to reduce our production of such irritants as industrial pollution. Such an anti-environmental stance is also apparent in the work of John Maddox, whose 1972 *The Doomsday*

Syndrome, while anticipating some of the arguments made here, naively advocates cultivating the entire Amazon basin and outrageously applauds the extinction of certain species (1972:86).

The Promethean *environmentalism* advocated here, by contrast, values progress as much for the benefits that it may confer to the human-ravished landscape as for promises it gives to the human community. Although it may seem preposterous, I agree simultaneously with Simon on the desirability of technological advance and with Dave Foreman (the archradical founder of Earth First!) on the necessity of wilderness restoration (see Foreman 1991:187). To understand such a seemingly paradoxical position, it is necessary to examine one of this work's fundamental theses: the belief that only by disengaging our economy from the natural world can we allow adequate space for nature itself.

The Decoupling of Humanity and Nature

The Promethean perspective adopted here advocates a form of environmental protection that green extremists would consider utterly heretical. Where they seek to reconnect humanity with nature, I counter that human society should strive to separate itself as much as possible from the natural world, a notion that has aptly been labeled "decoupling" by the geographer Simmons (1989:384). To advocate decoupling is to reject both the instrumentalist claim—that nature should be used merely for human ends—and the green counterargument—that humanity is, or should be, just another species in nature.

Decoupling processes have already averted ecological devastation many times. European forests, for example, avoided destruction when early modern smelters substituted coal for charcoal (see Perlin 1989). This process should continue as composites replace steel and as coal begins to yield to solar power—with nature breathing easier everywhere as a result. But one must wonder whether self-proclaimed deep ecologists affirming their communion with nature through shamanistic rituals will supply the world with solar technologies. I suspect rather that such delivery will come, if at all, from high-tech corporations—from firms operating in a social, economic, and technical milieu almost wholly removed from the intricate webs of the natural world.

If we are lucky, the commercialization of photovoltaic solar energy will come in good part from struggling American start-ups like Chronar. It now seems far more likely, however, that this technology will be dominated by such vast industrial concerns as Hitachi, Sanyo, and Fujitsu (*The Economist*, 19–25 May 1990). The engineers, investors, and managers of a company like Chronar should be hailed and supported as environmental heroes, not denounced as technocratic and capitalistic eco-villains. We will be better able to appreciate the vital roles that such companies play if we accept that ecological salvation will come through distancing ourselves from, rather than reimmersing ourselves in, the natural world.

To move from heresy to blasphemy, I would also suggest that as toxic waste decomposition technologies and recycling techniques are perfected, the use of synthetic materials will entail far less environmental destruction than will the continued production of natural products like paper, wood, and cotton. The future may yet be in plastics. Let us hope that companies like Du Pont can create artificial fibers sophisticated enough that we no longer need to deplete the earth's aquifers, clear its tropical forests, drain its wetlands, and pour massive quantities of biocides on all of these environments in order to grow the cotton that affluent American consumers consider so wonderfully "natural."

The greatest hope for virtually complete decoupling may lie in the so-called nanotechnology revolution (Drexler 1986; Drexler and Peterson 1991). If its proponents are correct, the nano techniques of molecular assembly will allow us to build superior goods using only a small fraction of the energy and materials now required. Indeed, Drexler goes so far as to argue that by mining surplus atmospheric carbon dioxide we will be able to provide most of the raw materials needed for the next economy. Moreover, not only would a nanotech economy spare the natural world of any noxious pollutants, but it would also allow a truly massive return of land to natural communities. Although the layperson may regard nanotechnology as utter fantasy, it is based on firm scientific reasoning, and it has been taken seriously by at least one prominent environmental philosopher (Milbrath 1989).

Nature for Nature's Sake—And Humanity for Humanity's
It is widely accepted that environmental thinkers can be divided into two camps: those who favor the preservation of nature for nature's sake, and those who wish only to maintain the environment as the necessary habitat of humankind (see Pepper 1989; O'Riordan 1989; W. Fox 1990). In the first group stand the green radicals, while the second supposedly consists of environmental reformers, also labeled "shallow ecologists." Radicals often pull no punches in assailing the members of the latter camp for their anthropocentrism, managerialism, and gutless accommodationism—to some, "shallow ecology" is "just a more efficient form of exploitation and oppression" (quoted in Nash 1989:202).

While this dichotomy may accurately depict some of the major approaches of the past, it is remarkably unhelpful for devising the kind of framework required for a truly effective environmental movement. It incorrectly assumes that those who adopt an anti-anthropocentric view (that is, one that accords intrinsic worth to nonhuman beings) will also embrace the larger political programs of radical environmentalism. Similarly, it portrays those who favor reforms within the political and economic structures of representative democracies as thereby excluding all nonhumans from the realm of moral consideration. Yet no convincing reasons are ever provided to show why these beliefs should necessarily be aligned in such a manner. (For an instructive discussion of the pitfalls of the anthropocentric versus nonanthropocentric dichotomy, see Norton 1987, chapter 11.)

In marked contrast, the decoupling perspective endorsed here seeks to separate human activities from nature *both* in order to protect nature from humanity (for nature's sake) and to allow continued technological progress (for humanity's sake). This entails acknowledging a profound division between humankind and the rest of nature, a distinction that many greens allege is itself at the root of the ecological crisis. Yet the radical environmentalists who condemn this example of dualistic thinking merely substitute for it their own parallel gulf, one separating modern (or technologically oriented) human beings from nature. This in turn entails positing a radical discontinuity in human development, a dualism of human nature separating moderns from primals (or primitives).

As I shall argue at length in this work's conclusion, such a division of humankind is, in the end, both bigoted and empirically unsupportable. We would be better off admitting that while humankind is indeed *of* nature, intrinsically creative human nature is a phenomenon not found in nature's other creations. In a Promethean environmental future, humans would *accentuate* the gulf that sets us apart from the rest of the natural world—precisely in order to preserve and enjoy nature at a somewhat distant remove. Our alternative is to continue to struggle within nature, and in so doing to distort its forms by our inescapably unnatural presence.

Finally, where radical greens often emphasize philosophical (or even spiritual) purity, this work stresses pragmatic gains. Since the anarchic utopianism that marks the dominant strains of radical environmentalism stands little chance of gaining public acceptance, much less of creating a feasible alternative economy, an emphasis on the purity of ideals can lead only to the frustration of goals. I would suggest that a pragmatic approach stands a much better chance of accomplishing our shared ends. The prospect of humankind someday coexisting easily with the earth's other inhabitants—a vision entertained by Arcadian and Promethean environmentalists alike—can best be achieved through gradual steps that remain on the track of technological progress.

Guided Capitalism

As noted above, I believe that only a capitalist economy can generate the resources necessary for the development of a technologically sophisticated, ecologically sustainable global economy. In embracing capitalism I do not thereby advocate the laissez-faire approach of the Republican right. To say that the market plays an essential role is not to say that it should be given full sway. As Robert Kuttner (1991) persuasively argues, the laissez-faire ideology has actually placed shackles on the American economy; it has rather been the "social market" economies, like that of Germany, that have shown the greatest dynamism in the postwar period. Moreover, if the example of Japan teaches us anything, it should be that economic success stems rather from "combining free markets and individual initiative *with social organization*" (Thurow 1985:60; emphasis added). At the same time, hard heads must always be matched with soft hearts (see Blinder 1987); we must never lose sight of social goals when working for economic efficiency or ecological stability. But both social equity and environmental protection are, I will argue, more easily realized by working through rather than fighting against the market system and the corporate structure of late twentieth-century capitalism. Economic growth, environmental protection, and social welfare should be seen as positively rather than negatively linked; a society that demands strict pollution controls, for example, will be advantaged in industrial competition at the highest levels of technological sophistication, as will a society that continually upgrades its human resources by providing workers with skilled, well-paying jobs (Porter 1990). It is not coincidental that Japan, seemingly poised to grasp world economic leadership, enjoys a much more equal distribution of wealth than does the United States—and a socialized medical system as well. The Japanese have never taken laissez-faire seriously (C. Johnson 1982), and if the United States further embraces it we will be sorely disadvantaged in the global economic race.

Nor should this work be construed as another manifesto for "technological optimism," a naive creed that environmentalists wisely disparage. We cannot blithely assume that unguided growth will solve our economic and environmental problems. But if we fail it will be in devoting too few of our resources to technology, not too many. More funds must be channeled into education, basic science, and long-term research and development if we are to find an environmentally sustainable mode of existence. While it is essential to guide technology into ecologically benign pathways, it is equally imperative that we consistently support the bases of technological progress itself.

A healthy society, I would argue, is one characterized by simultaneous increases in general prosperity, social equity, and environmental stability. The present trends are not encouraging; only a few societies are growing more prosperous, the gap between the rich and the poor is increasing both in the United States and in the world at large, and environmental systems throughout the planet are deteriorating. Yet we can devise ways to begin to even out social discrepancies and restore ecological health without sacrificing economic growth. I am convinced that such goals may be realized through "guided capitalism"—a corporate and market system in which the state mandates public goods, in which taxes are set both to level social disparities and to penalize environmental damage, and in which fiscal policies are manipulated to encourage long-term investments in both human and industrial capital (see Rosecrance 1990). But these social and environmental goals will, in the end, be attainable only if we nurture and guide rather than strangle the rather truculent capitalist goose that lays the golden eggs.

Uneconomic Despoliation
As we near the end of the twentieth century, the rallying cry of American radical environmentalism is the call to halt the clear-cutting of old-growth forests in the Pacific Northwest. On this issue, I side firmly with the eco-extremists. But while lauding their nonviolent holding actions, I by no means condemn the capitalist system that they hold responsible. While old-growth logging is often highly profitable to individual firms, it is utterly inconsequential to the American economy as a whole. Where ecological costs are great and economic benefits modest, we must opt for preserving nature. But even in strictly economic terms, if we were fully able to tally the long-term costs entailed by such reckless logging, we might well discover that the final balance sheet is in the red. It is time to implement a new economic calculus that fully encompasses environmental variables, one such as researchers at the World Resources Institute are presently refining (see, for example, Porter and Brown 1991:31).

In many cases subtle calculations are not needed so much as honest reporting. Much of the old-growth logging in the United States is economic insanity even in the short term. In the Tongass National Forest of southeast Alaska, the clear-cutting of ancient forests is profitable only because loggers are given massive subsidies. In fact, in the majority of America's national forests, logging is an uneconomic activity that would wither away without continual federal hand-outs (Repetto 1988). Most of the environmentally destructive water projects in the American West are similarly sustained from the public trough (Worster 1985), as is the massive overgrazing that occurs on almost all of the federal government's rangelands (Ferguson and Ferguson 1983; Libecap 1981). Even tropical deforestation is often stimulated by state subsidies and other governmentally mandated economic distortions (Repetto and Gillis 1988). In these cases avid proponents of capitalism ought to be just as incensed as the radical greens. That many self-proclaimed conservatives consistently support such subsidies may indicate that they care less about economic efficiency than they do for ensuring that key constituents are able to remain subsidized by the public purse.

In arguing against economically marginal and especially state-subsidized despoliation, one must recognize a hidden irony. Those who would clear the country's few remaining groves of uncut trees employ the same kind of romantic rhetoric commonly used by their antagonists. They argue, for example, that we are morally obligated to denude the slopes of southeast Alaska in order to save several hundred jobs and a few tiny communities. Most of the jobs will evaporate anyway, once the timber is depleted. But it is the rhetoric that is important; human communities come first, economic rationality be damned. Environmental protagonists and antagonists are thereby united in a mutual contempt for economics. It is in cases like this that hard heads are needed. If we embrace capitalism, we must accept that it necessarily involves what Schumpeter (1942) called "creative destruction." Old jobs will be lost, old communities will perish. A "soft-hearted" society would ease the transition of those persons affected, but it should be willing to let uneconomic—and intrinsically anti-ecological—industries fade into oblivion.

The ultimate irony of the late twentieth-century movement for environmental protection, however, is the fact that both the most strident greens and the most committed anti-environmentalists espouse a fundamentally similar—and thoroughly outdated—view of economic development. Both sides seem to believe that economic growth rests fundamentally on the exploitation of natural resources. The theory is the same, only the ethical positioning is reversed. Fortunately, the moral dilemma thus presented (should we opt for economic growth or environmental health?) is false, since the conceptual structure on which it rests is decrepit. Technologies, not natural resources, provide the essential motor of economic progress. If large segments of the American electorate continue to see forests and ore bodies, rather than research labs and product engineering centers, as the main repositories and wellsprings of national wealth, then we will seriously undercut our own well-being, and perhaps destroy the natural world in the very process.

While the claims of economic growth and environmental protection often do conflict, inefficiency in economic endeavors leads to its own severe forms of waste and degradation. Eventually we must realize, as economically oriented environmentalists are now telling us, that while the freedom to discharge wastes without cost into the environment may be a great boon to a given firm, it is remarkably destructive to the economy as a whole. Economic efficiency, at the most abstract level, is positively rather than negatively linked to most forms of environmental stewardship.

Political Pragmatism and Moderation

A sizable majority of American citizens are worried about the environment and are willing to sacrifice so that its health may improve. This is no simple partisan issue; Republicans and Democrats alike voice the same concerns (Borrelli 1988:19). But the entrenched ideological rifts cleaving American society prevent environmental issues from galvanizing adequate action. Moderate Republicans seek protection, but argue that market mechanisms provide the most efficient route. Many also call for more action within the private sector, pointing here to the remarkable successes of the Nature Conservancy (funded in good part through corporate donations). But moderates within the Republican Party have been severely marginalized in recent years, and it is becoming increasingly evident that the environmentalism of the Bush administration does not extend beyond campaign sloganeering. At the same time, radical environmentalists on the left dismiss the use of market mechanisms beforehand and deride the magnanimity of corporate donors as self-serving propaganda. Such thinking links ecological concern to a condemnation of business that the American public finds unacceptable. This intransigence, in turn, helps feed the far right within the Republican Party—the largest and most powerful group with an explicitly anti-environmental agenda.

If we are to preserve the earth, environmentalists must forge the broadest possible coalition. Major changes need to be made in public policy, changes that will require massive public support. That support can only be obtained by appealing to a centrist coalition. Yet at present, the large center ground of American voters, those who find merit in appeals both to economic efficiency and to social justice and environmental protection, is largely without an articulated platform. Party stalwarts, let alone

radicals, often regard moderates with contempt, viewing them as ideological weaklings unwilling to take a stand. I would argue the opposite. If
we are to take seriously the task of devising a sustainable future, it is
essential to admit that worthwhile ideas may be found on both sides of
this overdrawn political divide. As E. J. Dionne (1991:27) so brilliantly
argues, what is necessary is the creation of a new political center that
avoids "bland centrism" and instead seeks to build a genuine "coalition
for social reform."

Since critical theorists rightly point out that all writing is informed by
a political perspective, it is desirable to specify precisely the political
stance from which this work is composed. In simplest terms, I would
identify myself as a liberal moderate. The modifier "liberal" is apposite
because the great majority of the positions taken here, both explicitly
and implicitly, would be commonly classified as left of center. Moreover,
I fully concur with David Barash (1992) that the fundamental need is to
humanize capitalism, a project that he defines as the core of contemporary liberalism. But the essential term remains "moderate" because of
my insistence that dialogue and negotiation must be carried out across
the central divide of American ideology. In order to build an adequately
broad environmental consensus, we should endeavor to make that divide
as permeable to ideas as we possibly can.

It is especially important that environmentalists work with the leaders
of the largest corporations. Without corporate consent, a far-reaching
environmental reform program will prove chimerical. As will be discussed in chapter four, some companies have already made significant
environmental strides. Such firms are now working with moderate environmental groups, a process that has great potential if it is not undermined by eco-extremists.

To be sure, contemporary American leftist radicalism, in all of its
varied forms, exerts strong intellectual claims. Many of the thinkers
with whom I contend have come to a profound understanding of specific
problems and processes. Similarly, the visions they hold for a more just
future are rich and important. These thinkers must be taken seriously,
and I would not impugn the sincerity of their beliefs and actions. But one
must take equal care to avoid confusing moral outrage and sophisticated
dialectics with a legitimate claim to political power or with a desirable
(let alone possible) vision for humanity's future.

A Postmodern Environmentalism?
If the current environmental debate is fractionated, it is in part symp-
tomatic of the contemporary cultural condition identified as postmodernity. Separate "conversations," often marked by mutually incompre-

hensible vocabularies, are developing around disparate interest clusters. As this occurs, civil society, previously united in a coherent if divisive public discourse, threatens to dissolve. The postmodernist philosophical movement itself is a multifarious endeavor, united only in its distrust of monumental certainties, its experimental and often playful mood, its delight in irony, and its embracing of pluralism and diversity.

Conservatives, liberals, and orthodox marxists alike often find this postmodern mood highly disturbing, as it reduces their own roles from that of contenders for the political core to mere voices in a structureless cacophony. Certain environmental radicals, on the other hand, readily embrace postmodernism, finding in its attack on monumental structures and its celebration of pluralism a perfect vehicle for their own assault on the homogenizing force of global capitalism and their struggle to create diverse, autonomous, local communities (for example, Cheney 1989a; for an eco-radical denunciation of postmodernism, see Callicott 1990).

But the postmodern challenge can also be taken in an entirely different direction. One could argue that the notion of a single uniform "capitalism" that inevitably makes the world over in its own image is itself little more than a modernist myth (or, as Norman Cantor [1988] would insist, a neo-Victorian one). Capitalism, in fact, has proved remarkably protean, and its interaction with local cultures, while often destructive, can result in the emergence of diverse, hybrid sociocultural forms. Moreover, the influential art critic Charles Jencks (1987:36) contends that certain postmodernist impulses are actually at work creating a new unified public language of "free-style classicism." This, he argues, is the flexible and hybridizing tongue of a *common* urban industrial culture that wishes to draw on the stable heritage of the past just as it embraces the dynamism and the uncertainties of the future.

In important ways, this work adopts such a postmodernist perspective. While I seek to contribute to a broad public discourse, I do not call for conformity with a single political or philosophical stance. This work instead attempts to answer Stephen Toulmin's call for a return to a "practical philosophy" that dispenses with the grand search for "the universal, the general, and the timeless" (1990:186). Importantly, Toulmin upholds an ecological vision and, like the radical greens, he prefers egalitarian social arrangements to rigid hierarchies. Yet he also warns against blanket condemnations of such institutions as multinational corporations, seeking instead ways to ensure that they act responsibly (1990:207).

This vision strongly informs the critique presented here. The dominant school of radical environmentalism retains, despite its own intentions, a rationalist epistemology and a reductionistic mode of analysis. Its entire edifice rests on a few postulates, from which are generated a series of simple polarities: the wicked West versus the virtuous East, the evil industrials versus the peaceful primitives, and so on. Capitalism is reduced to a self-canceling impulse to accumulate for the sake of accumulation, while technical advance is assumed to be inextricably linked to an incessant concentration of power into ever fewer hands. Unfortunately for the greens, such rationalist figments vanish quickly in the light of empirical scrutiny.

Even if they long for a premodern world while touting the philosophies of postmodernism, most eco-radicals remain stubbornly modernist, at least in the broad sense of the term. In other words, they adhere to a philosophical tradition originating in the seventeenth century that seeks nothing less than absolute certainty (see Toulmin 1990). This unacknowledged philosophical foundationalism underlying the radical green critique of modern civilization is most tellingly exposed in the work of Arne Naess, founder of the deep ecology movement. Naess (1989; see also W. Fox 1990:260) expressly traces his philosophical lineage to the seventeenth-century thinker Spinoza—a pantheist, yes, but equally a Euclidean rationalist in the same vein as Descartes, whom most greens regard as the premier villain of intellectual history. While the two thinkers' views of nature may have been profoundly different, their modes of inquiry were deluded in exactly the same manner. Eco-radicalism, as presently constituted in its dominant school, has yet to escape this legacy.

In line with this work's particularist approach, I heartily encourage skepticism on the part of my readers—in regard to my own specific claims and suggestions as much as to those I criticize. For the new movement I envision, agreement is essential on two platforms only: that life on earth must survive, and that humanity itself should nonetheless thrive.

I will continue to argue, however, that a liberal-moderate stance offers the best hope for breaking the ideological impasse that currently paralyzes American society. Radical change *is* necessary, but I believe that it can come about only through concerted efforts to effect compromise and to seek broad-based conciliation. Unfortunately, the emergence of the social consensus needed to effect change appears unlikely. Ours seems to be an age of contention and separation, of sharpening ideological cleavages and widening socioeconomic divisions. Nor do many people seem adequately concerned over this state of affairs. Many conservatives are happy to see a growing rift between the rich and the poor, justifying it in the name of international competitiveness. And many radical leftists, for their part, revel in a culture of diversity in which it is assumed that people from different class or racial backgrounds are virtually unable to communicate, let alone agree. As long as such conditions prevail, ecological and economic catastrophe grows ever more likely.

I

The Varieties

of Radical

Environmentalism

■ The Five Contemporary Variants of Eco-Radicalism

Many eco-extremists argue that no core philosophy of any kind can be isolated in the radical green community, since the catchword of the movement is diversity (Manes 1990:21). While diversity does indeed characterize this literature, five main currents of thought can be discerned. By far the most influential is a soft variant of deep ecology, a philosophy perhaps best labeled "antihumanist anarchism." Similar to but distinct from this kinder and gentler school is a harsh variant of deep ecology that I call "primitivism"; primitivists advocate not merely the return to a small-scale social order proposed by other deep ecologists, but rather the active destruction of civilization. "Humanist eco-anarchism," the third major school of green radicalism, wishes on the contrary to perpetuate civilization's positive attributes, as does a fourth philosophy, "eco-marxism." Members of these latter two camps, however, disagree over whether the state or capitalism should be considered the ultimate force behind human destructiveness.

Cutting across each of these four schools is the broader philosophy of eco-feminism, which sees the ultimate origin of the rape of nature in men's exploitation of women. Eco-feminism comes in many forms, not all of which are radical. One extremist variant, however, so-called "radical eco-feminism," is of special relevance here. Radical eco-feminism offers a prescription for social change that is distinctly at odds with those of the other varieties of green extremism.

The following discussion outlines the tenets of the five schools of eco-radical thought identified above. For the first four—antihumanist anarchism, primitivism, humanist eco-anarchism, and eco-marxism—I attempt mainly to paraphrase their positions; criticisms will be offered in subsequent chapters. When the discussion turns to eco-feminism, however, the mode of inquiry perforce becomes more analytical; fine discriminations are necessary here to distinguish the doctrine of radical eco-feminism from its moderate cousin, so-called liberal eco-feminism, which this work strongly endorses. To conclude the chapter, the ideological basis of the predominant form of eco-radicalism is reexamined, with a view to situate it within the larger context of political philosophy. Here I will argue that much of radical environmentalism may be more accurately characterized as extremely conservative than as progressive in any meaningful way.

■ Deep Ecologies

The Soft Variant: Antihumanist Anarchism

Deep ecology preaches the gospel of "biospheric egalitarianism," which holds that all forms of life (in some cases, nonliving natural entities as well) have intrinsic worth and therefore an inherent right to exist, regardless of their potential utility for humanity.[1] Its advocates regard *Homo sapiens*, at least in its pretechnological manifestation, as an animal species just like any other, and they denounce all humanistic philosophies as inescapably arrogant and ultimately poisonous (see, for instance, Ehrenfeld 1978). Deep ecologists further regard nature as forming an interconnected totality, the whole of which is much greater than the sum of its parts. Consequently, true knowledge is held to be obtainable only through a holistic approach. The reductionistic methodology of Western science may isolate individual trees for invasive scrutiny, but it will be forever blinded to the deeper reality of forests.

Deep ecologists call for the human community to decentralize and to reintegrate itself with nature. Small-scale, local communities are urged to strive for self-sufficiency and participatory democracy, while individuals are encouraged to minimize their material consumption in order to live lightly on the earth. Total human equality and complete pacifism are propounded as the hallmarks of a future *stateless* human society that has made peace both with itself and with nature.

A related feature of deep ecology is its advocacy of a spiritual or transcendental path to environmental salvation. Many adherents espouse Asian or animistic faiths, finding in them an intuitive understanding of natural processes. Others look to psychology; Warwick Fox (1990), for example, seeks a new basis for environmentalism in Maslow's transpersonal psychology, while Naess (1989) regards self-realization as one of the movement's cornerstones.

Most eco-radicals view diversity of opinion as the vital counterpart in human thought to the diversity of species supposedly found in a mature ecosystem. Naess (1989), for instance, does not insist that other deep ecologists share his interest in self-realization (or, as he prefers it, "self-realization!"). He sees his belief system as but one of many possible "ecosophies," and urges others to develop their own. But the limits of acceptable ecosophies are in fact strictly circumscribed. While celebrating diversity and toleration, deep ecologists dismiss virtually the entire heritage of Western thought as not only morally bankrupt but as actively leading us to ecological destruction.

While deep ecology is usually considered a coherent if pluralistic philosophy, one stream diverges so much from the main channel as to deserve a distinct appellation. This "hard" branch of the movement is most accurately labeled "primitivism."

Primitivism

Primitivism, the philosophy of the most extreme eco-radicals, shares most of the ideas outlined above, but with one distinctive difference: it insistently glorifies rather than abjures violence. Viewing themselves as nature's warriors, primitivists seek to destroy civilization in order to bring about a global return to hunting and gathering, the supposed state of human grace. "Back to the Pleistocene" is thus a favored motto (Manes 1990:235). To this end, Howie Wolke, a cofounder of Earth First!, lauds violence as both natural and necessary for building a truly radical movement; his coconspirator Dave Foreman bluntly dismisses nonviolence as "a bunch of new-age crap" (quoted in Nash 1989:196).[2]

In its primitivist manifestation, radical environmentalism truly reveals the misanthropy of which it is so often accused. Whereas other deep ecologists believe in biospheric egalitarianism "in principle" (Naess 1989:28), primitivists take this doctrine at face value. Human beings and slimemolds, whales and spirochetes—we all are fundamentally the same, knots in the vast, gloriously interconnected web called nature. Saving humanity from extinction is accordingly no more important than saving the smallpox virus (see Nash 1989:85).

According to primitivist doctrine, if modern human beings are unique it is only in their capacity for destruction. Contemporary human society is regularly depicted as a pox on the body of the earth (Foreman 1991:57). Because of humanity's current propensity for evil, several well-known primitivists have proclaimed that they would "rather kill a man than a snake" (quoted in Nash 1989:153). The only way to exorcise this malignancy is for the vast bulk of humanity to disappear, with the few survivors renouncing all technological innovations made over the past 10,000 years.

The primitivist view is held only by a small segment of the radical environmental community. Its blatant misanthropy and glorification of violence do not sit well with most greens—including many who belong to Earth First! In fact, as Earth First! recently began to address social concerns and to move toward a more radical *cultural* stance, its cofounder, Dave Foreman, resigned in order to form a new, more hard-line organization (Gabriel 1990:64). Foreman, for his part, feared the movement had sacrificed its original vision, one founded on wilderness preservation, in favor of a vague revolt against the established social order: "More and more old-time EF!ers fail to attend the annual Round River Rendezvous because they feel it has transmogrified into a hippy/punk revel. Some leading Earth First! activists have turned from defending the wilderness to storming the barricades of capitalism. . . . The anarchist faction, frustrated with what they see as a stodgy *Earth First! Journal*, is now producing a 'punkzine' that aims to cultivate an impalpable 'wildness within'" (Foreman 1991:217).

■ Views from the Classical Left

While the primitivists attempt to pull the radical environmental movement away from social issues, other constituencies are tugging in the opposite direction. Influenced by the classical doctrines of the left, a small but growing group of activists calls for achieving ecological stability precisely by addressing human concerns. Whereas other eco-radicals are concerned above all with the human assault on nonhuman life forms, these thinkers are worried primarily about the effects of environmental deterioration on human communities. Two schools vie for influence here, each pulling leftward, but from different angles. On one side are eco-marxists, who argue that capitalism is the ultimate destructive force—one that can only be overcome through class struggle. Their opponents, the humanist eco-anarchists, stress rather the corrupting influence of social hierarchy, particularly as expressed in and sanctioned by the state. Let us begin with the latter ideology.

Humanistic Eco-Anarchism

American humanist eco-anarchism is dominated by Murray Bookchin and his school of "social ecology." Inspired by Kropotkin and other classical anarchists, Bookchin (1986; 1989) forwards a vision of decentralized, autonomous communities governed through participatory democracy and existing in full harmony with nature. Ecological health, Bookchin believes, is absolutely inseparable from social equality. Since social hierarchy is bolstered by precapitalist, capitalist, and communist countries alike, the state per se is abhorred as a manifestation of evil.

The humanism of social ecology is manifested in two complimentary ways. On the one hand, unlike deep ecologists, Bookchin privileges humanity over other species, while on the other, he retains a strong belief in human rationality. His vision is fundamentally socially progressive. While he finds a certain wisdom in pretechnological societies, Bookchin has no desire to return to the womb of nature. Instead, he espouses a kind of "small-scale technological optimism," believing that ingenious and environmentally benign technologies can render human want obsolete (Bookchin 1972).

While Bookchin has a devoted coterie, his ideas have had minimal impact on the radical environmental movement as a whole (W. Fox 1990:37). Although anarchism lies at the heart of deep ecology and has guided most European green parties, the explicit *humanism* of social ecology is repulsive to most American eco-radicals (Eckersley 1989). Nor has Bookchin influenced scholarly leftist environmentalism. Despite Bookchin's impressive command of political philosophy and his voluminous publication record, marxism rather than anarchism predominates in eco-radical constituencies found within the academic left.

Eco-Marxism

American eco-marxism is strictly an academic movement to date, albeit one with the potential to command considerable intellectual power.[3] Marxism in one form or another dominates several subfields in the academy, and some of the country's sharpest minds are committed to a marxist agenda. But orthodox marxism downplays environmental concerns, and an eco-marxist movement has been very slow to develop. Indeed, many marxists argued until recently that environmentalism was little more than a bourgeois diversion from society's real problems (see Pepper 1989:175–76), while even today many regard "green analysis" as hopelessly conservative (see Dobson 1990:31). Nonetheless, in the last decade, as a variety of heterodox marxisms have flowered, ecological issues have begun to appear on the agenda. With the launching of the new periodical, *Capitalism, Nature, Socialism*, eco-marxism has come of age.

Most marxist environmentalists swim against several main currents of eco-radical thought. Only a few self-styled "green-reds" share the widespread aversion to technology and centralized control; such fears, after all, run counter to the larger marxian heritage. So does the spirit of reverence for the natural world; as Bramwell (1989:33) bluntly puts it, Marx himself simply "did not like nature." Similarly, the common environmentalist's litany that the average American consumes too many of the earth's resources has been denounced by the doyen of eco-marxism, James O'Connor (1987:34), as a reactionary threat to the working class.

Eco-marxists also diverge markedly from most radical environmentalists in their assessment of the role of popular ideology. Deep ecologists often argue that the transformation of consciousness, accomplished through both pleas to reason and appeals to the spirit, is the key to averting ecological catastrophe (for example, Devall and Sessions 1985: 158). Marxists disparage such reasoning as naive idealism, seeing the root of the problem in sociopolitical structures rather than in the mental universes of individuals.

Many marxists go further, denying the significance of individual behaviors of any kind. This repudiation of individual responsibility may be philosophically trenchant, but it can also be highly convenient personally. Radical greens are often vocal about their sometimes intense personal guilt, realizing as they do that their lifestyles are never fully consonant with their ideals. Eco-marxists can dispense with this irritant, however, as a mere figment of bourgeois individualism. A marxist academic can thus command and spend an upper middle-class salary (a princely sum, by global standards) and still feel no sense of complicity in the ongoing destruction of the planet (for a novelist's deadly parody of such reasoning, see Lodge 1984).

These conflicts betray an underlying tension between marxism's Promethean vision and the Arcadian impulse of mainstream radical environmentalism, a discrepancy that has been the cause of fervent debate. Would-be green-reds have pored over Marx's and Engel's corpus trying to find some basis upon which to erect an alternative environmentalist theory, arguing vigorously among themselves over whether their varying interpretations are faithful to the tradition of dialectical materialism (see, for example, D. Lee 1980; Tolman 1981; D. Lee 1982; see also Jung 1983). Anti-marxists, in turn, have labored to show the futility of this endeavor by extracting anti-environmental sentiments from the same literature (Routley 1981; Clark 1989).

Ultimately, those involved in the attempt to devise a philosophical basis for eco-marxism have been forced to amend Marx's understanding of economics substantially. The essential innovation is to agree with the greens that, given the limits of a finite planet, economic growth cannot continue indefinitely, even under a communist regime (M. O'Connor 1989). Marxism, with this proviso, is able to affirm the centerpiece of radical environmentalism. What the green-red movement cannot endorse is the favored socioeconomic prescription of other green radicals: autarkic decentralization. Instead, eco-marxists hope to capture the eco-radical movement and convert its adherents; as Faber and O'Connor (1989:177) put it, the challenge is "to redden the mainstream Greens."

Eco-marxists might be expected to express concern if not embarrassment over the disastrous environmental records of all marxian regimes. Indeed, many do acknowledge the environmental failings of what they often gracelessly call "really existing socialism." But for most this remains a mere footnote to their real concern. Like other academic marxists, the green-reds devote almost all of their effort to analyzing the failings of capitalism; the less scholarly task of devising a blueprint for a better society receives surprisingly little attention. To these as to other marxist academics, capitalism is all-encompassing, and everything that matters is comprehensible only in relation to the vast webs of profit and exploitation woven by the global bourgeoisie.

Not surprisingly, most radical environmentalists continue explicitly to reject key elements of the marxist approach, noting correctly that Marx himself, in addition to being anthropocentric, supported political centralization and Western imperialism, while applauding virtually all forms of technological advance. Moreover, few eco-marxists express concern for the loss of biological diversity (except insofar as it might harm human communities), a stance not likely to gain much favor in a movement founded in large part on the principle of biospheric egalitarianism.

While eco-marxism thus remains a minor side stream, the sustained marxian critique of capitalism has deeply influenced the entire radical environmental movement. As chapter five will demonstrate, wittingly or not, most radical greens are substantially marxian in their understanding of political economics.

■ Eco-Feminism

The philosophy of eco-feminism cross-cuts—and seeks to transcend—the varied positions outlined above. Like the environmental movement as a whole, eco-feminism is fragmented into politically and philosophically disparate camps. What holds them together is the proposition that the destruction of nature is largely perpetrated by men and follows logically from their exploitation of women. Ultimately, eco-feminists argue, a single masculine "logic of domination" (K. Warren 1990:132) lurks under all forms of oppression and exploitation.

Not all adherents of eco-feminism are radical. Liberal eco-feminists, who stress individual freedom within the framework of capitalist democracy, are by definition moderate. In eco-feminist circles, however, liberalism has few adherents. Socialist (and more specifically, marxist) eco-feminism, which places class exploitation alongside gender oppression as a central category of analysis, is more widespread. But even socialist views are secondary within the broader stream of feminist environmentalism, whose dominant school has been ambiguously christened "radical eco-feminism." While the marxian eco-sisterhood's *politics* are more conventionally radical, radical eco-feminists counter that their *feminism* is the more uncompromising of the two (on the varieties of eco-feminism, see Thrupp 1989; K. Warren 1987).

The Doctrines of Radical Eco-Feminism
Radical eco-feminism rests on an essentialist proposition: the notion that men and women are different in their essences, that they intrinsically think and feel differently (Salleh 1983; see also K. Warren 1987: 13–15). Most importantly, women are considered by nature closer to the natural world than are men. Radical eco-feminists blame the male mode of apprehending reality for our present ecological and social ills. Masculinist ideology is said to be guilty of three fallacies: it glorifies the atomistic individual, it posits hierarchies of distinction leading inevitably to naked structures of domination, and it is based on false "normative" dualisms separating people from nature and mind from matter. To avert crisis and restore both the earth and humanity, we must embrace a female, compassionate, nonhierarchical, nondualistic, ecological mode of thought structured around relationships rather than individual entities (Kheel 1985; K. Warren 1987).

Radical eco-feminism is closely linked with earth-based spirituality. Many adherents are actively reviving the goddess-centered cults that they believe once allowed humans to live in harmony with nature. Several leaders have proclaimed themselves witches and formed covens devoted to the ancient faith of Wicca (Starhawk 1989). Inspired by the revisionist writings of such archaeologists as Marija Gimbutas (1980), they argue that European societies were peaceful, nonstratified, harmonious, and goddess-worshipping until the invasion of patriarchal, militaristic, Indo-European pastoralists in the fourth millennium B.C.E. With the dismantling of patriarchy and capitalism and the reinstitution of goddess-centered spiritual practices, radical eco-feminists believe that such a social order could be reinvented.

Radical eco-feminism shares much of its philosophy with soft deep ecology. Both schools emphasize biospheric egalitarianism, decentralization, participatory democracy, pacifism, spiritual (or at least psychological) union with nature, and the eschewing of all "hard" technologies. Yet a deep gulf of suspicion separates the two groups. Eco-feminist writers of various persuasions have denounced deep ecology as merely another convoluted pretext for continued male domination—an accusation that finds its mark with the hard or primitivist version, if less readily fitting the dominant, anti-humanist anarchist school (see Salleh 1983; Cheney 1987; Doubiago 1989).

For their part, deep ecologists argue that, as abhorrent as patriarchy may be, the exploitation of nature by humanity is the ultimate problem. Warwick Fox (1989) goes further in countering that eco-feminism simplistically reduces environmental destruction to a single factor and in so doing fails to apply ecological thinking. Thus, despite their similarities, deep ecology and radical eco-feminism remain fundamentally unreconciled.

Feminist Challenges to Radical Eco-Feminism

Radical eco-feminism has also been challenged by mainstream feminists. Especially contentious is the issue of essentialism. To many feminist thinkers, the notion that female thought is fundamentally different from male thought in any way—especially by being more in tune with nature—is both offensive and dangerous. This very line of reasoning, they point out, has long been used by men to keep women in subjugation (see K. Warren 1987:14,15; Plumwood 1988). When women are identified with nature, men are able to appropriate the realms of science and rationality for themselves. While radical eco-feminists may condemn such forms of thought as masculinist, their efficacy—both for progress and for exploitation—is not thereby canceled. By spurning rather than seizing the tools of science and technology, radical eco-feminism itself may unwittingly help perpetuate patriarchy.

In more pragmatic ways as well, radical eco-feminism and, to a lesser extent, marxist eco-feminism have profoundly antifeminist implications in practice. The former movement advises women to turn away from existing means of wielding public power. Since large-scale institutions are, by definition, irredeemably patriarchal and exploitative, women are called away from existing positions of public power (Plant 1989:187). Instead, all feminists (men as well as women) are enjoined to retreat into separatist, autonomous communities. Marxist eco-feminists do not demand such hermetic exclusion, but their philosophy too calls ultimately for struggle against rather than participation within capitalist society. Since institutional science, corporations, and large public institutions are, despite radicals' fondest hopes, well entrenched, such withdrawal risks disempowering women still further. A refusal to seek positions in such imperfect institutions as presently exist would relegate women to the role of sideline critics, undermining their opportunity to be participants—and indeed leaders—in the ongoing restructuring of society.

In its effort to avoid the appearance of cultural imperialism, radical eco-feminism also flirts with an ethical relativism that could conceivably undermine the feminist agenda at the global scale. To posit that "[w]hat counts as sexism, racism, or classism may vary cross-culturally" (K. Warren 1990:139) is to ignore a huge array of deeply sexist practices existing in numerous non-Western cultures.

Finally, the successful realization of the radical eco-feminist dream would threaten women in a very immediate sense. In the anarchic world they envision, men—who are certainly more physically powerful than women and appear to be more inclined toward violence as well—could easily arrogate power at the local level and devise neo-patriarchies. Anarchists argue that humankind's inherent good would prevent this—a view accessible only to those wearing the deepest of psychological blinders. As will be shown in chapter three, many primal societies, contrary to eco-romantic fantasies, were unabashedly patriarchal.

Liberal eco-feminism, considered by most to be the movement's weakest and most compromised wing, in my view offers the best promise for achieving both environmental and feminist goals. In fact, as will be argued at some length in chapter six, a liberal eco-feminist approach is not merely desirable but is rather essential if we are to create an ecologically sustainable global social order. The stabilization of world population alone is a feat that will require women to be massively empowered. As long as women are held in patriarchal bondage, as they literally are in countries like Pakistan, fertility rates are likely to remain unsustainably high.

■ Eco-Radical Politics: Neither Left nor Right?

The conventional notion of a one dimensional political spectrum, pitting a progressive left against a conservative right, fails to capture the nature of contemporary eco-radical discourse. Indeed, many radical environmentalists proudly proclaim that their philosophy is neither left nor right (Milbrath 1989:132). The dominant version of green extremism, in fact, does share basic characteristics with both radical leftism and ultraconservatism. Partly because of this ideological twist, journalist Virginia Postrel (1990) suggests that the most fundamental rift in American politics today is not that separating the left from the right, but rather that lying between those who favor economic growth and those who prefer "green" stasis.

Ghosts of the Radical Right

Yet the majority of contemporary radical environmentalists would consider themselves leftists of some variety. To be sure, only a small minority are classical leftists, as most explicitly reject all long-established socialist doctrines as guiding philosophies. In advocating communalism and denouncing capitalism, however, even deep ecology fits comfortably into the larger tradition of left-wing thought.

But while the green movement as a whole is almost always classified by outsiders as "left," several prominent environmentalists have espoused distinctly right-wing views. Not coincidentally, a few have even been vilified as eco-fascists. Garrett Hardin (1968; 1977), for example, not only calls for the full privatization of common property resources in order to prevent a global "tragedy of the commons," but has actually suggested mass starvation as a cure for overpopulation. Equally generous is the position of those Earth First! members who hopefully look to AIDS as nature's antibody to the human plague (Conner 1987).

While these frightening sentiments characterize only the far fringe, eco-radicalism shows its illiberalism in surprisingly numerous forms. Owing to their commitment to total decentralization, for instance, many radical greens denounce central intervention even in local communities that choose to dispense with such notions as "democracy, equality, liberty, freedom, justice, and the like" (Sale 1985:108). (This is, of course, the essential argument of the deeply conservative states' rights campaigns of earlier generations.) Robert Heilbroner (1980:172–73), on the other hand, has argued that the transition period leading to the development of a small-scale social order will probably require centralized social organization of a highly disciplined, quasi-military, monastic form. Still other eco-radicals advocate restrictions on present-day freedoms at the national level; many would like to curtail reproductive choice sharply (Ehrlich and Ehrlich 1990), while some would be happy to gut the First Amendment, especially as it pertains to advertising (Catton 1980:236).

The connection between ecological rhetoric and reactionary politics is a fact of long standing. Throughout the past century, as Anna Bramwell (1989) persuasively demonstrates, ecologism has been linked as often with the far right as with the far left. British naturists in the interwar period, for example, commonly defined themselves as "High Tories" (Bramwell 1989:51). More disturbingly, German national socialism was heavily laden with a romantic, ecologically imbued "blood and soil" ideology. Modern-day American nazis are similarly injecting an environmental element into their racism; as the notorious Tom Metzger haltingly argues, "increasing numbers of young people in the white racialist movement are also quite interested in the ecology . . . , and it seems to me that we are becoming more aware of our precarious state, the white man, the white woman's, state in the world, being only about 10 percent of the population, we begin to sympathize, empathize more, with the wolves and other animals" (quoted in Langer 1990:86).

Mainline green radicalism is also paradoxically conservative, albeit in a more gentle manner. While generally viewing themselves as either firmly on the left or as transcending the left-right political spectrum altogether, many eco-extremists look back fondly to ultraconservative critics of the industrial revolution such as Edmund Burke. As a group they also have much in common with the reactionary southern American agrarians of the mid-twentieth century (see Dionne 1991:154–55). Although eco-radicals deny the ultraconservatives' contention that class divisions are an expression of natural hierarchies, they embrace the latter's more general—and wholly reactionary—premise that society can and should conform to nature's rules. Thus, environmental radicals like Young (1990) urge modern greens to forge an alliance with the few remaining ultrapaleoconservatives (those who reject Adam Smith as dangerously liberal) based on a shared longing for stability, tradition, and family.

Although eco-radicals universally denounce class divisions, William Tucker (1982) has argued powerfully that environmentalism as a whole reeks of aristocratic conservatism. The upper middle-class "suburban agrarian" backbone of the ecology movement, he claims, is motivated primarily by a desire to protect its own playgrounds and parkways from poorer people. According to Tucker, environmentalism's aristocratic heritage is visible principally in the desire of comfortable suburbanites to maintain their own habitats by "preventing others from climbing the ladder behind them" (1982:15).

Tucker's argument fits with the views of certain mainstream environmentalists of the 1970s better than with those of the eco-radicals of the

1980s and 1990s. Moreover, his case is overstated and his text reveals frighteningly anti-environmental sentiments. Yet his main point must be taken seriously. The implementation of green radicalism's political agenda would severely reduce the possibilities for social mobility. Such a freezing of class lines is, after all, the prime desiderata of aristocracies everywhere. While the greens would counter this tendency by advocating massive redistribution (see Kassiola 1990:72), one must realize just how difficult it is to achieve thoroughgoing redistribution and to maintain income equality over time; even marxian communism, it is now evident, has certainly failed at the task.

One reason conservative ideas so readily latch onto the environmental discourse lies in the widespread denial of any essential distinction between human beings and nature's other creatures. If people are fundamentally similar to all other animals, it is only logical that our basic nature should derive from instinct and other biological imperatives—a staple of conservative thought for centuries. In contrast to this organicist view is not only the notion that human nature is at least substantially a cultural construct, a central tenet of contemporary social theory, but also ideas of individual autonomy and free will—the core of most classical justifications for civil liberty. Both derive from a distinctly different *humanistic* tradition. Civil rights are much more difficult to defend in a discourse that purports to explain human nature in biological terms.

Eco-radicals reject the accusation of biological determinism, and indeed many of their writings fiercely deny the doctrine. But it is a profound contradiction to insist, on the one hand, that humans are (or should be) a species in nature like any other, while maintaining, at the same time, that human nature is intrinsically cultural rather than biological. One way out of this dilemma is to counter that animal personalities are also culturally constructed (see W. Fox 1990:117)—a proposition not entirely unreasonable in regard to apes and perhaps certain other warm-blooded animals (Degler 1991:344), but utterly ludicrous when considering hookworms, sponges, jellyfish, and indeed the vast majority of nonhuman species.

A similar threat to human liberty is implicit in the holistic view of nature advocated by the dominant school of eco-radicalism. To a thoroughgoing holist, individuals are inconsequential "moments in [the] network, knots in this web of life" (Callicott 1986:310). Because it places the good of the whole far above that of the mere individual, holism has been denounced as totalitarian and even fascist by several noted environmental thinkers (Kheel 1985; see also Marietta 1988).

Finally, even the religious orientation of many eco-radicals reveals a potentially reactionary bent. Zen Buddhism, in particular, has been embraced by many radical environmentalists (W. Fox 1990:70–71), who evidently feel that through Zen meditation they may experience communion with the natural world. The antirational practices of Zen, however, have historically proven highly susceptible to abuse by political and military elites seeking to gain unconditional obedience from their subordinates. As van Wolferen (1989:316) claims: "Actually, the historical function of Japanese Zen, which thrived among the warrior class, was to lower the resistance of the individual against the blind obedience expected of him, as can be gathered from the common Zen imagery of 'destroying' or 'extinguishing' the mind." Indeed, all of the Asian creeds so eagerly embraced by eco-radicals have been associated with notoriously antiliberal political regimes.

The Commonalities of Extremist Thought

All forms of political extremism, be they identified as right or left, unite in their disdain for democracy as we know it. While many lavish praise on a hypothetical form of direct or participatory democracy (for example, Tokar 1987), all implicitly disparage representative democracy as a sham, a circus staged by big-money interests. (Some would go so far as to suggest that "representation" is fraudulent as long as animals and plants have no parliamentary voice [see Nash 1989:130].) Given the fact that the electorate almost always rejects their call, this attitude is hardly surprising. In return, the more extreme eco-radicals regard the majority of voters with barely concealed contempt, viewing them as blind stooges easily deceived by the blandishments of a corporate-owned mass media (see, for example, Worster 1985:57–58).

If green extremism rejects representative democracy in theory, its response to existing American governmental institutions is one of extraordinary contempt. Most eco-radicals would like to demolish our entire political foundation, and many argue that all national-level structures are inherently destructive of the environment. Here they can find much agreement on the far right. Indeed, a newly touted extreme right-wing version of environmental philosophy, the so-called "free-market environmentalism" of Anderson and Leal (1991), is as hostile to the American government as is the environmentalism of the far left. The only difference is that rightist radicals would like to see power devolve onto wealthy individuals and corporations rather than the properly constituted communities or politically acceptable social classes favored by the far left.

A third commonality of the extremist mode of thought is the thesis

that social health can be restored once some foreign, corrupting influence is extirpated. Naturist ideology, in both its left and right incarnations, has usually identified the alien force disconnecting humanity
from the redeeming natural world as capitalism. Whereas nazis attacked
the Jewish community in part because of its supposedly intrinsic capitalistic nature, the contemporary eco-radical movement lays special
blame on urban white male business executives. But the pattern is the
same. Ultimately, political extremism's scapegoat-seeking mentality
persists in all forms of eco-radicalism (again, see Bramwell 1989).

■ The Varieties of Eco-Radicalism Reconsidered

Each of the five schools of thought outlined above—antihumanist anarchism, primitivism, humanist eco-anarchism, green marxism, and radical eco-feminism—is presently battling for the heart of the radical environmental movement. While the first now occupies the seemingly
stable mainstream, the larger flow seethes with intellectual turmoil.
The preceding pages have admittedly caricatured these various schools,
although some writers are so extreme that even parody is all but impossible. More importantly, the division between the various camps is
not always as clear-cut as has been suggested above. Many individual
writers readily draw theoretical elements from a number of disparate
philosophies.

Such cross-fertilization is, however, not always mutual. While several
influential deep ecologists, for example, have praised the humanistic
eco-anarchist stance (Devall and Sessions 1985:18), Bookchin has hardly
returned the compliment; in his appraisal, deep ecologists are "well-to-
do people who have been raised on a spiritual diet of eastern cults mixed
with Hollywood and Disneyland fantasies" (1989:11). Another humanistic eco-anarchist has linked the deep ecology movement to "fascism,
genocide, corporate power, and, worst of all, bad epistemology" (quoted
in Manes 1990:155). Ironically, it appears to be the anti-humanists,
rather than the caustic humanists such as Bookchin, who are the more
humane in practice.

Many other green thinkers, including some of those best known to the
public, espouse philosophies that combine radical and moderate perspectives. Barry Commoner (1990), for example, hopes for a "genuine socialism" and in places veers toward anarchism, but he appears decidedly
moderate next to Bookchin or James O'Connor. The ecological economist Herman Daly, for his part, valiantly attempts to inject a degree of
economic rationality into the radical greens' communitarian, low tech-

nology, no-growth vision (Daly 1977; Daly and Cobb 1989). Paul Ehrlich holds radical positions on a few issues, but takes moderate stances on many others. And finally, John Dryzek (1987) argues powerfully for a "rational ecology" that avoids many of the pitfalls of extremism, yet in the end he dismisses the market for being antisocial while credulously claiming that in anarchy "authority can stem from the possession of a better argument" (1987:216). (For other prominent views that skirt the terrain between radical and moderate environmentalism, see Kassiola 1990 and Milbrath 1989.)

These eclectic voices, however, are beginning to be drowned out by those of the true extremists, writers who, like Christopher Manes (1990), call for nothing less than the total demolition of civilization. Those who are convinced that humanity is destroying nature generally advocate taking the strongest medicine conceivable. Unfortunately, they are proposing a lethal overdose.

"Ecologism," to borrow Bramwell's words, has long been "a convincing box in which all kinds of alternative ideas and people fit," despite its many apparent paradoxes (1989:237, 249). Yet while the wide range of ideologies presented here demonstrates that no single unitary philosophy of radical environmentalism can be identified, most green extremists do nevertheless share a critical core of beliefs. As I see it, these include the following propositions: the notion of primal purity, the imperative of total decentralization, the necessity of implementing appropriate technology, and, in the final analysis, the belief that capitalism itself is a fundamental evil. As each camp inflects these propositions differently, however, the many variations on these four standard themes provide ample material for sustained analysis. Let us begin with the notion that by following the primal way we can return to a harmonious relationship with the natural world.

2

Primal Purity and

Natural Balance

■ The Radical Position

The belief that rural peoples with simple technology offer a model of environmental stability and social sanity runs deep in eco-radical thought. Only such peoples are thought to be truly connected with nature, a supposed prerequisite for a healthy existence. Before the Second World War, European ecological romantics revered peasant culture and often looked back to the Middle Ages as an environmental idyll (see Bramwell 1989); American environmentalists have similarly long upheld North American Indians as ecological exemplars. Recently, scholars like Booth and Jacobs (1990) have further claimed that Native American ideology affirms the arguments now propounded by deep ecologists and eco-feminists. Other contemporary greens have globalized this vision, finding environmental wisdom in all rural societies but especially among the tribal (or primal) peoples least corrupted by technology, markets, and state-level political structures.

In order to affirm the thesis of primal purity, eco-radicals must recuperate not just tribal society but also nature itself. "Humanity in nature" is not an appealing prospect if the latter is pictured as crimson-toothed. Eco-radicals have thus embraced a distinct ecological theory to accompany their anthropological vision, one that glorifies the harmonious functioning of undisturbed ecosystems and that considers cooperation among individuals and among species far more common than competition.

Varieties of Eco-Radical Interpretation
The varied schools of radical environmentalism view primal societies through different lenses. Humanistic eco-anarchists are selective, lauding only specific tribal attributes. Bookchin (1989:47, 50), for example, praises precapitalist societies for being strikingly nondomineering and basically in tune with nature, but he considers their parochial fear of outsiders a fatal flaw. Green marxists similarly dismiss the notion that tribalism might offer a way of life superior to that of industrial civilization.

Deep ecologists, by contrast, lavishly praise tribal lifeways, social covenants, and ritual actions. Adherents of the dominant school of antihumanist anarchism often caution that seeking "inspiration from primal traditions" need not imply advocating a return to "the stone age" (Devall and Sessions 1985:97), but the latter is precisely what the more hard-core primitivists desire. As the editor of *Earth First!* recently proclaimed, "many of us . . . would like to see human beings live much more the way they did fifteen thousand years ago as opposed to what we see now" (quoted in Manes 1990:237).

Just how far back we must go to discover true harmony is a matter of debate in eco-radical circles. Writers aligned with European green parties often focus on traditional peasant communities. Brian Tokar, for example, finds medieval social organization highly admirable. "Right through the Middle Ages, peasant villages in Europe were generally free of the pressures to overproduce . . . and thus were able to devote nearly as much time to festivals and religious celebrations as they did to their own survival needs" (Tokar 1987:11). Primitivists, on the other hand, are inclined to see humanity's downfall as beginning much earlier, often with neolithic agriculture (Manes 1990:29). One green radical's gentle cultivators are thus another's eco-thugs, relentlessly attacking nature by spreading mutant monstrosities in the form of domesticated plants and animals.

The Supposed Superiority of the Primal Way
The primal tradition is touted as superior to that of modern society primarily because it allowed people to live "in sustainable communities for tens of thousands of years without impairing the viability of ecosystems" (Devall and Sessions 1985:127). "For most of human history," radical greens believe, "people lived in close harmony with the natural world" (Tokar 1987:9). Primal peoples formed an intrinsic part of nature, their lifeways finely tuned to its rhythms and their subsistence patterns intricately balanced with those of other species.

The nature invoked here is typically one whose original state is envisioned as virtually static, exemplifying equilibrium. Individual elements of nature may undergo change, but the whole—the marvelously intricate network of interdependencies—is believed to have persisted in perfect harmony, unaltered in its essence. Even the introduction of *Homo sapiens* did not at first disturb this web. For millennia, the natural world encompassed primal humanity as just another species, thus maintaining balance. But when human beings discovered technology, unity was rent. A single species now separated itself from, and in the process began to destroy, the rest of nature.

Most eco-radicals believe that humanity's long period of grace as a species in nature is reflected internally in primal social organization. As Tokar (1987:9) writes, "nature itself formed the basis for human cooperation and human freedom." Evil, in this view, stems from the breach between people and nature, of which tribal societies are innocent. Primal peoples are culturally healthy, unsullied by social hierarchies and uncorrupted by personal greed for material objects. In a striking reversal of the classical Eurocentric view, many eco-radicals go so far as to aver that only tribal peoples have developed true democracy, one in which every member of the community freely participates in an ongoing affirmation of social and natural unity.

This harmony of the whole is held to be most strikingly evident in religion. As Plant (1989:113) writes, "tribal peoples [see] the spiritual as alive in us, where spirit and matter, mind and body, are all part of the same living organism." Primal rituals affirm the profound connectedness of nature. When they must kill, tribal hunters redeem the act spiritually by asking forgiveness from the prey's spirit. Despite human predatory behavior, the intrinsic psychic harmony spanning all of creation can thus persist undamaged.

Radical environmentalists sometimes go further to outline an eco-primal mode of thought, often explored under the rubric of Native American consciousness. Such consciousness was purportedly based on a direct or unmediated apprehension of reality, leading to both social and ecological stability. Carolyn Merchant (1989:20) articulates this view in unequivocal terms:

> For native American cultures, consciousness was an integration of all of the senses with the body in sustaining life. In this mimetic consciousness, culture was transferred intergenerationally through imitation in song, myth, dance, sport, gathering, hunting, and planting. Oral-aural transmission of tribal knowledge through myth and transactions between animals, Indians, and neighboring tribes pro-

duced sustainable relations between the human and the non-human worlds. The primal gaze of locking eyes between the hunter and the hunted initiated the moment of ordained killing when the animal gave itself up so that the Indian could survive. . . . For Indians engaged in an intimate survival relationship with nature, sight, smell, sound, taste, and touch were of equal importance, integrated together in a total participatory consciousness.

Many radical environmentalists further argue that only in such a primal setting can a fully human existence be realized. In the modern world, individuals are alienated from each other through hierarchy and competition; in the tribal realm, they are united by cooperation and mutual care. In addition, members of small-scale, rural societies are believed to excel in the finer arts such as drama (usually enacted in rituals), art, music, craftwork, and so on. Unburdened by false desires for superfluous material objects and resting ever assured in the love of kith and kin, people in the primal realm are truly fulfilled. It is precisely their harmonious coexistence within nature—their constant contact with the wild plants and animals around them—that fills them with a supreme satisfaction that the rest of us have forgotten. This Garden of Eden once covered the entire globe; today it persists only in a few remote and threatened preserves.

■ Scholarly Inspiration

The primal impulse has long pushed against the march of civilization, often as a vague longing for Arcadian serenity, but occasionally as an openly articulated challenge to the status quo. With the development of an explicitly ecological critique of modern society in the late nineteenth century, opponents of industrialization increasingly found inspiration and justification from science itself. Ecology and ethology showed that human beings were merely another animal species, while so-called energetic economics indicated that industrial progress was unsustainable, given the finite supply of resources (Bramwell 1989).

With the vigorous renaissance of radical environmentalism in the late 1960s, the neo-tribal, countercultural impulse inspired new scholarly movements. Although the larger countercultural wave soon ebbed, its basic tenets survived in small but intellectually prominent groups such as the present-day radical greens. More broadly, a general distaste for modern life suffused many branches of American academia. The following pages explore how the 1960s' idealizations of primal humanity and of nature's all-encompassing balance were reflected in, and reinforced by, scholarly accounts of tribal peoples and ecological systems.

The View from Ecology

Ecology as a scholarly endeavor was defined in the mid-1800s, by the German zoologist Ernst Haeckel, as the study of "nature's household." Early ecologists ambitiously promulgated not only a distinct methodology befitting a new field of study, but also a distinct metaphysics. Essentially, Haeckel strove to deny the Aristotelian dualism separating humanity from nature, hoping to replace it with a monistic conception, one envisioning a unitary world comprised of structurally bound elements (Zimmerer forthcoming). This notion has remained central to ecologically inspired philosophy to this day.

As befits their object of study, most ecologists since the earliest days have championed the preservation of nature. Recently, the term ecology has even come to connote in some quarters a philosophical, moral, and political viewpoint rather than a specialized field of scientific inquiry (for example, as used by Bramwell 1989). Although ecology as a discipline and ecology as a political philosophy are very different things, the former has contributed directly to the development of the latter.

For one thing, ecological science traditionally stressed connections and interrelationships above individual entities, encouraging the holism characteristic of eco-radical thought. Further, until recent years, ecological orthodoxy posited stability and harmonious interconnectedness as the fundamental characteristics of (undisturbed) ecosystems. A key concept to emerge was that of the "ecological climax," implying that a mature ecosystem existed in a virtually static state of equilibrium. "Perhaps the paramount law in ecological theory," writes the environmental philosopher Holmes Rolston (1989:14), "is that of homeostasis."

While the more sophisticated eco-philosophers warn against transposing the "is" of nature to the "ought" of human conduct (see Callicott 1982a), many radical greens have based their models of ideal social organization on ecological science. In doing so they have been forced to simplify greatly the insights of the ecologists. Haeckel had coined the phrase "competitive cooperation" to describe the double-edged nature of organism interactions in an ecosystem, and this slogan was taken up by a variety of nonradical, ecologically inspired social theories (see Zimmerer forthcoming). But modern-day eco-radicals have nearly written competition out of the equation, singling out cooperation alone as nature's standard mode. It is this move that has given rise to the notion of humanity's primal harmony existing within the manifold webs of a perfectly harmonious natural world. Many radical greens have clearly decided that cooperation is nature's signal characteristic; as Sale (1985: 113) puts it, "symbiosis is essentially the way nature has normally seen fit to work."

Many other supposed characteristics of nature's household have also been transposed into social prescriptions for a future green world. Andrew Dobson, for example (1990:24, 25), argues that since "stability in an ecosystem is a function of the diversity in that ecosystem," a stable social system would be marked by a diversity of opinions. He adds that "an ecosystem that is subject to fluctuation has not reached the 'climax phase' and is therefore characterized as immature," implying that mature ecosystems—and by extension mature social orders—will be characterized by stasis (1990:25). Dobson takes this line of thought in some rather improbable directions. Having identified "longevity" as another "principle feature of the natural world," he argues that "tradition" in human societies should be enshrined as exemplifying nothing less than ecological wisdom (1990:25).

Anthropological Contributions

A second source of scholarly support for the idea of primal purity has been anthropology. Since its inception under the guidance of Franz Boas, American anthropology had devoted itself to showing that so-called primitive peoples are in no way inferior to moderns (Boas 1940; Degler 1991). All peoples, ethnographers correctly argued, possess the same basic mental endowment. Indeed, countless studies demonstrated that technologically simple cultures often boast the most sophisticated conceptions of natural processes (for example, Conklin 1954; Levi-Strauss 1966; Johannes 1981).

Prior to the 1960s, however, anthropologists seldom viewed tribalism as the ideal form of social organization. While some argued that we might learn from primitives how to be more spontaneous or less sexually repressive, few advocated return. But when ethnological inquiry began to merge with ecological concerns, primal peoples came to be viewed increasingly as environmental exemplars.

In the 1950s a group of anthropologists delineated the subfield of cultural ecology, whose central tenet was that a society's "core" cultural forms are structured around its adaptation to the local environment (Steward 1955). Early practitioners retained a strong separation between culture, which they saw as based on cooperation, and ecology, which they contrastingly pictured as founded in competition (Ellen 1982:64). This distinction was to become blurred in the 1960s and 1970s, a period that saw the flowering of a broader-based, cultural-ecological approach in anthropology. The new "general ecology," or "ecosystemic," perspective conceived of human-environmental relations as unified within a single "monistic" system (Ellen 1982:75). The human groupings appropriate for field study were now defined as "populations"—reproductive units essentially comparable to those of animal species—rather than as "cultures," the distinctly human congregations seen as fundamental by earlier ecological anthropologists (see Rappaport 1979:58).

Despite these departures, eco-systemic anthropologists continued to focus on harmonious relations between human groups and their physical environments. Their catchword was adaptation, a concept borrowed from evolutionary biology. Practitioners sought to demonstrate how specific cultural features, ranging from basic agricultural practices to elaborate rituals, functioned to ensure that a group's relation with its environment would remain in perpetual balance (for example, Rappaport 1967).

"Environmental adaptation" had for some time been used as an explanatory device in theories of cultural evolution, but only in the late 1960s did the idea acquire moral overtones. Anthropologists soon began to argue not only that tribal societies were intricately fitted to their environments, but that industrial societies were, in contrast, grossly maladapted (Rappaport 1979). Eco-salvation, it now seemed to many, could only come from heeding tribal ecological wisdom.

Inspired scholars quickly discovered specific material ways in which tribal economies demonstrated their superiority. "Primitive" horticulturalists grew diverse swidden (or slash-and-burn) gardens, allowing rapid forest regeneration by mimicking the structure of tropical rainforests (Geertz 1963). Modern farmers, in contrast, engaged in monocropping that could only be supported by massively destructive chemical subsidies. Moreover, whereas swidden cultivators were able to harvest many more calories than they expended in cultivation, modern farmers consistently operated at a net energy loss, clearly visible once one tallied the energy consumed by their machines and embodied in their chemical inputs (Rappaport 1971).

At a more abstract level, ecological anthropologists theorized that tribal social organization was also intrinsically adaptive, particularly in its resilience in the face of disturbances. A local disaster might completely destroy a given community, but it would leave more distant villages—those not immediately in the path of destruction—unscathed. Modern society, by contrast, was described as "hyper-coherent"; even a localized catastrophe could generate severe malfunctions over far-flung regions linked closely through trade (Rappaport 1979:162). Industrial civilization was thus seen as hovering on the brink of a precipice, vulnerable to being pushed over the edge by even a minor disruption.

An important source of inspiration for ecosystemic anthropology in this period was cybernetics, the relatively new science of control systems and automata. This might seem incongruous, given the movement's naturalist bias, but a number of commonalities united the two philosophies. As articulated by Gregory Bateson (1972), the cybernetic vision—like the ambitious general systems theory in which it was grounded—posited a fundamental unity of all phenomena. Material objects, systems theorists suggested, could best be understood as nodes through which processes operated over vast systems. An ecosystem, as a collection of elements connected through energy exchange, could be regarded as structurally analogous to a computer network. It was the flow itself that ultimately counted, not the nature of the objects through which the flow momentarily surged. Not surprisingly, the emblem of systems theorists came to be the flow chart, a diagram designed to specify precisely the sequence and magnitude of such exchanges.

Bateson's notion of systemic unity was strikingly similar to the radical monism of Haeckel and other nineteenth-century ecologists. In a famous metaphor, Bateson (1972:458; 1979:210) argued that it was no longer useful to think of a man engaged in cutting down a tree as an independent entity. Instead, the man, the axe, and the tree should be seen as temporarily forming an integrated circuit, united by flows of both energy and information—an emergent system, as it were. From here he led the reader along a series of steps to what were in fact rather dramatic theological speculations, concluding, with Haeckel, that deity should be considered not transcendent but immanent, embodied within the phenomena of nature. "The individual mind is immanent but not only in the body. It is immanent also in the pathways and messages outside the body; and there is a larger Mind of which the individual mind is only a sub-system. The larger mind is comparable to God . . . but it is still immanent in the total interconnected social system and planetary ecology" (Bateson 1972:461).

This pantheistic apprehension of nature has remained a touchstone of most varieties of radical environmentalism (for example, H. Wood 1985; McKibben 1989:71). The fascination with cybernetics, on the other hand, although briefly fashionable among appropriate technology fans and readers of Co-Evolution Quarterly, has since been discarded by most eco-radicals as irredeemably technophilic (Devall and Sessions 1985: 151). Many continue, nonetheless, to rely on the anthropological studies of the period that remain steeped in cybernetic discourse (see, for instance, Young's [1990] use of Rappaport [1967]).

Evidence for Primal Prosperity
Even if it could be shown that primal peoples lived in perfect harmony with nature, there remained the thorny issue of their living standards. Tribal peoples could hardly be upheld as models if their lives were as impoverished as popularly imagined. Here again, anthropologists came to the rescue. The public, it was now claimed, had misunderstood the true nature of tribal society; the uninitiated had inappropriately transferred their own parochial notions of prosperity to peoples with entirely different standards. The noted anthropologist Marshall Sahlins (1972) went so far as to argue that non-Western societies were, as a rule, unconcerned with anything so mundane as material wealth. As he quipped, "[m]oney is to the West what kinship is to the Rest" (1976:216). By the late 1960s and early 1970s studies were available to show that it was precisely the most primitive peoples, the hunter-gatherers, who existed in a state of true affluence (see Lee and DeVore 1968 on the !Kung San of the Kalahari, and Blainey 1975 on the Australian Aborigines).

Primal affluence, as notable anthropologists informed the public, was possible not because tribal peoples possessed abundant goods but rather because their wants were minimal (Sahlins 1972). In particular, modern hunter-gatherers like the !Kung enjoyed a great abundance of leisure. They typically "worked" only several hours a day—if one could even employ such a pejorative term to denote the pleasant tasks of gathering mongongo nuts and stalking game. Other studies indicated that swidden agriculturalists as well led idyllic lives, their needs abundantly provided by their nature-mimicking garden plots (Bergman 1980). The contrast could hardly have been more marked. While moderns toiled at miserable jobs to accumulate spiritually empty and earth-destroying trinkets, primal peoples devoted only a few hours a day to diverse and pleasant subsistence tasks, freeing up uncounted hours for socializing, merry-making, and communing with earth spirits.

A final rehabilitatory task was to show that, on top of everything else, primal peoples enjoyed genuine social harmony. Here the vast files of anthropological evidence were unfortunately mixed. For every group of gentle people, one could cite another that seemed to exhibit true savagery, the infamous Ik of Uganda being the most striking example (Turnbull 1972). Studies of contemporary hunter-gatherers, however, generally suggested that harmonious cooperation was at least their norm. Then, in 1972 a completely untouched stone-age tribe was purportedly discovered in the southern Philippines. In widely publicized reports, excited journalists and anthropologists described the Tasaday as primal humanity in its essence, totally isolated and completely nonaggressive (Nance 1975). This seemed to clinch the case that gentle cooperation was indeed the *original* human condition—and one that could be reclaimed if only we could summon the necessary will.

Finding evidence for primal sexual egalitarianism was more problematic. Anthropologists had long known that many tribal groups were unambiguously dominated by men. But again, new studies showed that cooperation between the sexes was a prominent feature at least in contemporary hunter-gatherer societies (R. Lee 1979:250). Even more important was the new evidence suggesting that European neolithic society was strikingly egalitarian, being neither socially stratified nor male dominated (Gimbutas 1980). Eco-feminists could now argue that women and men had once lived together in harmony, giving rise to hopes that they could do so again if freed from institutions of domination.

Cultural Geography's Ambivalent Contributions

The anthropologists seeking to prove the superiority of tribal ways were joined by a smaller cohort from geography. Cultural geographers had long studied the relationship between human beings and their natural environments, and many had argued, long before it became fashionable, that technologically less-advanced societies tended to use the natural world in a much less destructive manner than did their industrial counterparts (for example, Sauer 1938 [1963]). This early work in cultural geography reinforced the idealization of the primitive.

At the same time, the historical perspective of the discipline undermined the notion that primal peoples were changelessly and harmoniously adapted to the natural world. Geographers were especially interested in the human transformation of the environment, demonstrating how natural landscapes were turned into cultural landscapes through even the most basic subsistence activities (Thomas 1956; Goudie 1981; Simmons 1989). Where anthropologists discerned adaptation *to* nature, geographers had been trained to notice purposive modification *of* nature. Wherever human beings could be found, they insisted, pristine nature was an anachronistic concept.

In addition, those cultural geographers trained in the so-called Berkeley school had long based their inquiries on an analytical separation of humanity from nature (Zimmerer forthcoming). Upholding the notion of "culture" as an autonomous phenomenon, these scholars rejected the idea that human society could be analyzed according to either mechanistic or biological principles. In marked contrast to ecosystemic anthropologists, cultural geographers insisted that any human society must be approached through a detailed empirical exploration of its historical development. To be sure, the Berkeley school was not without analytical shortcomings; some critics have even accused its members of reifying culture as a "superorganic" principle of organization (Duncan 1980). Yet cultural geography continues to offer a clear precedent for the empirical and historical turn that I will use below in challenging the romanticization of the primitive.

Recent Trends in Ecological Anthropology and Geography
The reinvention of the myth of the noble savage by late twentieth-century scholars was not without its merits. Popular images of tribal peoples living brutal lives on the bare edge of subsistence were simply fraudulent. But in their rush to devise a more relevant and generous approach to the classical subject of anthropological research, not a few scholars were blinded by their own desire to rediscover Eden.

Although romantic notions popularized in the early 1970s have retained currency in a few quarters (for example, Wirsing 1985), most anthropologists have over the past fifteen years rejected the central propositions of neoromanticism outright. The idea that tribal peoples are so finely attuned to their environments as to exist in a state of timeless homeostasis has been roundly criticized and essentially abandoned. Marxists scholars, in particular, have successfully argued that no societies are either changeless or free of internal conflict. At the same time, leftist social theorists have denounced the larger project of ecosystemic anthropology for being predicated on the inherently reactionary philosophy of functionalism (the belief that sociocultural traits survive because they function to maintain the status quo) (Friedman 1974; see also Ellen 1982:89–92). The rigid adaptationist paradigm also proved unable to accommodate the proliferating evidence that sociocultural transformations seldom mirror environmental change (Bargatzky 1984), just as it was severely criticized for leaving scant room for human thought and creativity.

In recent years ecologically informed anthropological research has pushed forward in several different guises, but its main proponents have firmly turned away from romanticization. Indeed, many prominent practitioners have sought a hard-headed scientific respectability that would horrify the deep ecologists who continue to find inspiration in their predecessors' work. Jochim's (1981) influential *Strategies for Survival*, for example, employs a model of human rationality derived from behavioral psychology, neo-classical economics, and evolutionary biology. Allen Johnson (1982:416), meanwhile, has argued that the very distinction between reductionistic and holistic modes of analysis is useless, claiming that "one man's reductionism is another's holism, depending on where each stands in the chain of inclusiveness from basic particle elementism to universal oneness." The radical academic impulse, for its part, has introduced the unsympathetic tradition of marxian theory, which has a long history not only of stressing social conflict over harmony but also of decrying the "idiocy of rural life." By the 1980s many anthropologists and cultural geographers had concluded that the most important issue to explore in regard to indigenous peoples was not their traditional adaptation to nature but rather their current destruction at the hands of industrial society. Inspired in part by the marxist critique, the realization began to dawn that many tribal groups faced imminent extinction.

This process too is now seen to have an inescapable ecological dimension. As numerous studies have shown, once formerly independent groups become enmeshed in larger economic circuits, even the most ecologically adaptive practices often collapse. Geographers have, for example, documented dramatic increases in soil erosion and deforestation in areas recently marked by more harmonious environmental practices (Grossman 1984; Blaikie 1985, Blaikie and Brookfield 1987). Other studies have examined the devastation that occurs when remote tribal territories are invaded by external economic interests bent on extracting resources for their own benefit (for example, Hecht and Cockburn 1989). By the late 1980s this new school of "political ecology" had captured the imagination of many scholars working on the margins of the world economy.

Defending the rights of indigenous peoples to their own lands and resources—a platform fully endorsed by Promethean environmentalism—is, however, a very different matter from idealizing them as existing in a state of grace. To argue that indigenous societies should have the right to determine their own developmental pathways does not mean that we must emulate their traditional ways of life. Indeed, many tribal peoples today strongly wish to participate in global circuits of commercial exchange, and when they are afforded adequate protection from outsiders they are often able to modernize their economies selectively without sacrificing their cultural identities (see, for example, Fürer-Haimendorf 1982, chapter 11). To deny them such opportunities by relegating them to "cultural zoos"—the stance implicitly taken by some eco-radicals—is both untenable and reprehensible.

Furthermore, the time has come to admit that so-called primal peoples were not as a rule fully adapted to their environments and did not exemplify true social harmony. Their societies, like those organized at a larger scale, have proved capable of the full range of human good and evil. Yes, they should be defended and, in most cases, admired. But as the following pages will attest, they simply cannot serve as models for humanity's future.

■ Eden Revisited: Scholarly Miscues

The eco-radical thesis of primal purity falters on both empirical and theoretical grounds. Beginning with the latter failing, we shall see that many of the theoretical constructs employed by environmental romanticists obscure the terrain well before any evidence can be assessed.

Over-Generalizing the Primal
Most importantly, radical environmentalism's vision of primal harmony is so exaggerated as to verge on intellectual fraud. Its fundamental error is overgeneralization. Eco-radicals seldom distinguish among tribal groups, picturing them instead as the undifferentiated "other." The result is that, despite acknowledgment of surficial diversity, primal human-environmental relations are seen as being essentially the same everywhere. "For the *primal mind* there is no break between humans and the rest of nature" (Devall and Sessions 1985:97; emphasis added). The deification of the earth, Sale (1985:5) informs us, "appears with such regularity . . . in every preliterate culture, that we may think of it as a basic, almost innate, human perception."

A specific example of unwarranted overgeneralization may been seen in the work of Christopher Manes. Following Michael Roselle, Manes argues that all tribal peoples possess an environmentally sensitive legal tradition that one may call either Pleistocene law or *adat* (the latter term borrowed from the Penem of Borneo). "Although tribal peoples may have widely different forms of *Adat*, they apparently all share the affirmation that humans are part of a larger ecological community toward which they have certain responsibilities" (Manes (1990:173). Manes engages here in an extraordinary act of reductionism. Without any exploration of the evidence, a vast array of human cultures that bear little resemblance to one another are collapsed into a single, unified category. In fact, *adat* is a specific concept applicable only to certain Malayan peoples. It is by no means universal within Southeast Asia, much less to tribal society at large. Nor do all tribal societies actually affirm human responsibility to the larger ecological community. To do justice to the variety of human cultures, each group must be approached as a unique entity, not as a reflection of what we think we have lost.

A similar problem mars the work of Carolyn Merchant (1989), who writes of native American consciousness in the singular. Evidently, Merchant believes that habits of mind were fundamentally uniform across the tremendous array of societies inhabiting pre-Columbian North America. Throughout native American society, we are told, knowledge was "monistic" and consciousness "mimetic"; in contrast, in industrial society knowledge is "dualistic" and consciousness "analytic" (1989:25).

Merchant's conception of native American consciousness, precariously based on the ruminations of a handful of scholars, is ethnographically unsupportable, if not outrageous. The kind of evidence necessary for making such generalizations does not exist—in fact, it never could exist. Consciousness varies tremendously, both from individual to individual and from culture group to culture group. (As anthropologist Vincent Crapanzano [1986:74] pithily demands, "How can a whole people share a single subjectivity?") Reductionism on this scale cannot help but distort one's findings. For example, Merchant (1989:24) informs us that American Indian populations existed in steady-state equilibrium, which is patently untrue (see below). Moreover, the notion that analytic consciousness is a monopoly of Europeans has long been a favorite of racist ideologues. Although Merchant would have us reverse the moral signs and hold the mimetic consciousness of native Americans superior, the very distinction remains invidious.

Anthropologist Marshall Sahlins (1976:216) invents an equally simplistic moral universe, one in which societies of the West are structured around money, while the rest are organized around kinship. It would seem that the premodern urban civilizations and trade networks of Asia have been erased from Sahlins's historical map altogether. Or perhaps we are being asked to believe that they were held together exclusively by kin relations, without the cement of commercial bonds. Either alternative suggests a remarkable blindness toward non-Western history. As Anthony Reid (1988) has clearly shown, indigenous states and tribal groups in Southeast Asia have in fact been complexly integrated for centuries through indigenous commercial circuits; the literature attesting to similar features of Chinese and Japanese history would fill a library.

The tendency to overgeneralize is so strong that even scholars who recognize its pitfalls often stumble into them anyway. This is strikingly evident in Callicott's work (1982b) on American Indian attitudes toward nature. Although he begins by warning that there was "no one thing that can be called *the* American Indian belief system" (1982b:293), Callicott proceeds to construct just such an ideological scheme based on a handful of examples. He supports this move by claiming that "Europeans appear to be a more ethnically diverse, motley collection of folks than Indians" (1982b:295), a statement that is left unsupported because it is unsupportable. Most of Europe was culturally integrated via Christianity in the medieval period, whereas pre-Columbian North America lacked any sort of integrative cultural structure.

Ecological Distortions

As noted above, the postulate of primal purity is closely bound with the notion of the mature ecosystem as a diverse and relatively changeless assemblage of organisms. Recent scholarship has largely discredited this vision. The perfectly balanced climax ecosystem is now seen as a chimera, a reflection of social ideals rather than of nature's actual workings. Modern ecological theory stresses not homeostasis but continual flux (Pickett and White 1985). Nor does contemporary ecology accept the neat correlation between diversity and stability (Connell 1978; Norton 1987:60); in fact, as Brennan (1988:104) demonstrates, "natural forests will sometimes display diversity precisely because they are disturbed by factors such as fire, weather damage and human manipulation." Such harmonies as do exist in nature, ecologist Daniel Botkin (1990) powerfully argues, are best seen as discordant.

Most eco-radicals seem oblivious to this theoretical revolution in ecology. And while the more sophisticated among them have followed the changing scientific currents (for example, Worster 1990), few seem to realize their implications for green social theory.

Many practicing ecologists, however, have begun to reassess the standard environmentalist view of the proper connection between people and nature. No longer, scientists like Botkin argue, can we dream of fitting passively within a preexisting balance. Not only do natural communities continually change, but the human impact itself is inescapably transforming. The best we hope for is to minimize our deleterious effects through wise management. In fact, as Botkin (1990) demonstrates, in trying to remove all taints of humanity from a given area, in a hopeless attempt to restore primordial conditions, we often only accelerate processes of degradation.

If many radical greens have not heard this message and remain wedded to outdated ecological constructs, others have apparently never bothered to master the basic principles of the science in the first place. Thus one encounters such fantastic assertions as: "The drawdown of forests . . . has already disrupted the vital 'breathing' functions of worldwide photosynthesis and may have overshot the earth's ability to restore enough carbon monoxide to the atmosphere" (Sale 1985:30). The "green lung" thesis implied here has been proved erroneous; in a mature forest as much oxygen is consumed in decomposition as is released in photosynthesis. The expressed fear of carbon monoxide depletion is, on the other hand, simply bizarre.

Ecological theory similarly suffers when used by eco-radicals to justify their social programs. Even the more scientifically literate writers frequently misapply ecological principles to human societies, reifying what can only be poor metaphors into specious explanations. Catton (1980: 133), for example, argues that unemployment can be usefully understood as an example of "niche saturation." To analyze what is, in fact, a political and economic phenomenon in terms of an ecological construct—which even in its original context verges on tautology—is to obscure the conflictive, interest-based nature of the interactions that perpetuate it. Workers are not species, and we can never hope to understand the complex social origins of a problem like unemployment by resorting to the lexicon of ecosystem dynamics.

Finally, the radical's model of ecosystem dynamics is itself seriously flawed. The prevailing notion in the eco-extremist literature is that cooperation is nature's standard mode of interaction. To be sure, many examples of cooperation can be found, but when two individuals belonging to different species cooperate, it is often because they are both keen to avoid being devoured by a third. Another common occasion for cross-species mutual aid occurs when one animal consumes parasites that cooperatively feast on another. Even within a single species, competitive and sometimes extremely vicious behaviors are as commonplace as gentle acts of caring. Think of postcourtship behavior in black widow spiders, or of the battles between male elephant seals at breeding season. In the latter case, not only do bulls maul each other in pursuit of dominance, but they not uncommonly trample their own pups to death in the process. In fact, many ethologists now argue that even animal behaviors that appear to be altruistic are ultimately self-interested (see Degler 1991:281–85).

In short, the eco-radical view of both natural communities and primal peoples clearly falls short of the mark on conceptual grounds, succumbing to massive overgeneralization and betrayed by faulty and outdated theoretical constructs. The remainder of this chapter is devoted to a systematic marshaling of empirical evidence to illustrate the inadequacy of the resulting views. The historical record suggests that in many cases the harmony that modern scholars claim to see exists more in their imaginations than in reality.

■ Primal Discord

The present task is to challenge the notion that people in tribal societies invariably live in harmony and ease with their environments and with their fellow human beings. Many small-scale societies have indeed exemplified social concord and ecological sustainability, but generalizations cannot be drawn from a few isolated cases. For every example of a group living harmoniously one may cite another in which discord seems paramount. In fact, tribal peoples are often highly destructive of nature; their lives are, in most cases, more penurious than affluent; and as a rule they exhibit the same range of social pathologies that may be found within industrial societies.

Radical environmentalists, however, will continue to forward vast generalizations based on one or two case studies. Often the examples of harmonious existence that they cite are valid. (A cautious stance, however, is always warranted, since preconceptions about the primal condition may unconsciously distort ethnographic findings.) If, contrarily, I cite numerous cases of staggering environmental devastation, it is not to argue that destructiveness is therefore the tribal norm. Some small-scale societies have lived in relative harmony with nature while others have despoiled their landscapes in drastic fashion; similarly, while some have been gentle and peaceful, others have been astoundingly brutal. Unfortunately, due to the polemic nature of this work, I must concentrate here on the negative instances.

Primal Pillaging? Pleistocene Extinctions

Primitivists, the most extreme eco-radicals, argue that as soon as plants and animals were domesticated true primal harmony began to vanish. Yet even hunter-gatherers have been guilty of environmental despoliation. In fact, much indirect evidence suggests that roughly 11,000 years ago paleolithic hunters perpetrated the earth's most horrific human-induced ecological tragedy: the extermination of most large mammals in North and South America. Let us begin, therefore, in the Pleistocene epoch.

Some 11,000 years ago, a brief interlude in geological terms, the Pleistocene Ice Age came to an end. With it vanished approximately 85 percent of all large mammals in North America. These extinctions were part of a global wave of species death that struck with greatest severity on the peripheral continents of North America, South America, and Australia. Eurasia was less seriously affected, Africa least of all. African extinctions occurred at the earliest date, American extinctions significantly later, and extinctions on remote islands most recently (Martin and Klein 1984).

The earth has witnessed many other episodes of mass extinction, but the Pleistocene die-off was unique in several respects. Its geographical patterns were curiously discontinuous, but more unusual was its general restriction to large mammals and, to a lesser extent, large birds. Mammalian megafauna on the hard-hit continents was, however, devastated. Major evolutionary lines, such as that of the ground sloths, perished entirely. As appalling as the extinctions of plants and arthropods currently occurring in tropical rainforests is, it has not yet matched the ecological destruction that occurred when several continents' largest and most widespread animal species perished.

Since the end of the Pleistocene, North America has been a faunal wasteland. Our mammalian diversity should equal that of Africa—as it recently did. A host of large mammals had easily survived the ebb and flow of glacial and interglacial climates over the Pleistocene's many hundred thousand years. Were is not for this ecological holocaust, mammoths and mastodons, giant ground sloths and gargantuan armadillos, saber-toothed tigers and dire wolves, American camels and American horses, giant beavers and short-faced bears, and many other species as well, would have greeted the first Europeans to land on this continent.

North America will never be the same. Mammals and plants had coevolved over millions of years, and when the creatures vanished, plant ecology was dramatically altered as well. We know, for example, that African elephants significantly transform their landscapes; in large numbers they can turn a forest into a savannah or even a treeless grassland. We should well wonder what pristine interglacial vegetation would have been like on a continent that boasted two families of proboscideans—elephants (mammoths) and elephant-like (mastodons) animals—as well three families of giant ground sloths, some species of which were of elephantine bulk.

Ironically, most of the few American survivors of this disaster, animals such as bison, bighorn sheep, and the large deer (elk, moose, and caribou), were relatively recent immigrants from Eurasia. They are not, in other words, evolutionarily genuine Americans. This distinction belongs largely to the large mammals—horses, camels, and sloths—that perished.

Scholars have feverishly debated the cause of Pleistocene extinctions for many years. One school blames climatic change; an opposing camp argues for overkill by paleolithic hunters. Neither side can prove its case definitively, and it is possible that humans may someday be proved completely innocent. The majority of specialists, however, support a multicausal explanation that holds human beings at least partially responsible (Anderson 1984:41).

Even if proponents of climatic causation are substantially correct, the eco-radical notion that stability is a hallmark of natural ecosystems still loses all credibility. Writers like Guiliday (1984) and Graham and Lundelius (1984) strongly contend that familiar modern-day American plant communities and even broad-scale biomes (such as coniferous forests, deciduous forests, and grasslands) simply did not exist in the recent past. Pleistocene landscapes, they claim, were rather characterized by complex, patchy, vegetational mosaics that contained within them species that today do not exist within several hundred or even several thousand miles of each other. Much of North America, the evidence suggests, had been something like a spruce forest, but also a bit like a grassland, and yet still somewhat reminiscent of a sagebrush steppe.

According to the environmental models proposed by these writers, climate change some 11,000 years ago shattered these complex Pleistocene plant communities. With a switch to more pronounced seasonality, the plaid or mosaic form of the past was replaced by the broad zones, or stripes, that characterize the vegetational patterns of modern-day North America. Climate change advocates suggest that large mammals, which had evolved in the patchwork world, either could not keep pace with these transformations or simply could not find an adequate variety of habitats to support themselves in the newly simplified landscapes.

Clearly, these scholars have shown that modern-day North American biomes are grossly impoverished when compared to those of the Pleistocene. As Graham and Lundelius (1984:243) conclude: "The high density and low diversity characteristic of many modern-day plant associations such as the spruce forest may be explained by the disruption of co-evolved systems." In this view, the relative stability of an ecosystem such as the boreal spruce forest may ironically be more an artifact of its *low* diversity—especially the absence of that quintessential spruce browser, the mastodon—than a result of its supposedly *high* diversity.

Yet while the increased seasonality at the end of the Pleistocene is well documented, and while in all likelihood it did exert a profound stress on large mammals, to my mind climate change alone cannot account for megafaunal extinctions. South Asia and Southern Africa are now, and have likely always been, as seasonal as central Mexico or the central Brazilian plateau, yet *massive* extinctions occurred in the latter and not in the former regions. North Africa, which retained elephants until well into historical times (witness Hannibal's feat), has a close climatic analogue in California—which was stripped of its proboscideans along with virtually everything else. Moreover, many extinct American forms, such as the horse, thrived when they were reintroduced to appropriate landscapes on this continent, just as they had continued to thrive in similar Asian climatic regions until historical times.[1] Such divergent conditions in similar environments may well point to human agency. Even scholars who argue most doggedly for climatic change often concede that paleo-Indian hunters may have played an important secondary role (Guiliday 1984:250, 257; Guthrie 1984:290).

Other evidence strongly supports the human overkill hypothesis as well. Paul Martin (1984; Martin and Wright 1967), its principle proponent, convincingly elucidates the geographical patterns of extinction in terms of human evolution and migration. In Africa, he argues, large-mammal hunting developed gradually, allowing humans and their prey the opportunity to develop stable patterns of coexistence. But as technologically adept big-game hunters migrated to distant continents, the resident large mammals, unexposed to human ways, were easily exterminated along an advancing wave of human settlement. The large mammals that survived the disaster tended to be "weedy" species with fast reproductive rates, such as various extant members of the deer family (McDonald 1984).

Strong evidence for the overkill scenario also comes from the few areas that humans did not reach in the Pleistocene. In Europe, for example, many species survived for a period on Mediterranean islands that remained inaccessible to *Homo sapiens*. "Ironically the last European elephants appear to have been dwarfs occupying oceanic islands, an environment inevitably viewed by biogeographers as especially prone to the hazards of natural extinctions" (P. Martin 1984:390). The evidence is even more clear for Madagascar and New Zealand. On those islands, large animals persisted until human beings arrived some 1,000 to 2,000 years ago, at which point massive extinctions ensued. Moreover, waves of species death followed Polynesian seafarers not just to New Zealand but to other Pacific island groups as well, most notably Hawaii. Finally, the last of the Pleistocene extinctions, that of the giant Steller's sea cow, did not occur until the eighteenth century, when its remote, unpeopled refuge in the Bering Sea was finally discovered by Russian sailors. In the Pleistocene epoch, this gentle, easily killed marine herbivore had been widespread in coastal waters as far south as California (on island extinctions in general, see the various essays in Martin and Klein 1984).

Nonspecialists often dismiss the human-agency thesis out of hand. They do so, I believe, not because they can refute its arguments, but rather because it contradicts their cherished myths about primitive peoples. Few radical environmentalists have begun to realize the extent of their error in continuing to imagine that until the advent of Europeans the North American landscape had existed in a harmonious and static balance.

No matter how convincing the evidence, the overkill hypothesis remains just a hypothesis. In order to fully dispel the myth of primal purity, it is necessary to consider more recent and more thoroughly documented cases of primal pillaging. Remembering the primitivists' suspicion of agricultural societies, we will begin by considering peoples who rely on hunting and gathering for all, or at least most, of their sustenance.

Primal Wildlife Destruction in Historical Time

Severe overhunting by so-called primal peoples has occurred in many different parts of the world. As it turns out, not only do hunter-gatherers (and hunter-farmers) sometimes slaughter their prey in a nonsustainable manner, but in several instances they have actually been encouraged to do so by their religious ideology.

One of the clearest examples of anti-environmental actions and sentiments among a tribal population may be found in Robert Brightman's account of an Algonquian group, the Rock Cree, inhabiting a section of North America's boreal forest. According to Brightman (1987:123), the Rock Cree historically not only lacked a conservation ethos, but evidently had a "proclivity to kill animals indiscriminately in number beyond what was needed for exchange or domestic use." In fact, their religious consideration of animal spirits led not to a conserving respect for nature, but rather to utterly unnecessary destruction. Evidently, members of this group believed that "game animals killed by hunters [would] spontaneously regenerate after death or reincarnate as fetal animals"; in other words, the more animals were killed, the more the species as a whole would increase (1987:131). As one eighteenth-century witness wrote, "they were so accustomed to kill everything that came within their reach, that few of them could pass a small bird's nest, without slaying the young ones, or destroying the eggs" (quoted in Brightman 1987:124).

Eventually, Brightman demonstrates, the Rock Cree did adopt conservation practices at least for fur trapping, but the impetus came largely from the Hudson's Bay Company. Company traders knew that they would lose their business if the indiscriminate slaughter continued, giving them firm motivation to press for sustainable yields. Conservation eventually became incorporated into the indigenous religious system, but with an unusual twist; many Rock Cree now came to believe that humans could decimate animal populations, but "only if the animals [or their spirits'] *permit* hunting and trapping to have that effect" (Brightman 1987:137).

Other studies similarly suggest that not all hunting peoples understand the long-term consequences of overharvesting. The Shoshone Indians, for example, once "sought to obliterate the local animal population in the interest of maximizing the immediate food supply" (Johnson and Earle 1987:34). Hames (1987:103) writes of Amazonian hunters who "regard protein as sufficiently valuable to intensify their efforts for it in the face of [animal] depletion." Carrier (1987:153) found that on Ponam Island near New Guinea, religious ideology again prevented the development of a conservation ethos. Villagers here considered the impact of human actions within nature to be negligible, since real power exists only in the supernatural sphere. The extinction of a given species was seen as an act of divine will, and as such "not necessarily undesirable" (Carrier 1987:153).

Hunters lacking a conservation ideology sometimes began to slaughter wildlife with abandon upon obtaining modern weapons. Ornstein and Ehrlich, for example, describe an Eskimo seal hunt in which "something on the order of twenty seals killed by rifle fire sank out of reach in the fresh water for every one retrieved" (1989:59). The authors conclude that Eskimos generally reject the tenets of conservation, again because they believe that animal spirits, not population dynamics, control the future supply of prey.

Eco-radicals might argue that these are distorted examples, since the peoples in question had already been corrupted by Western ways. Truly primal peoples, by this line of reasoning, are those who have not yet been "contacted." Such pristine groups, of course, are conveniently inaccessible to scholarly inquiry. But even if one accepts such logic, it seems odd that hunter-gatherers who had lived affluent lives amid harmonious surroundings would suddenly become avaricious exploiters of nature as soon as they had an opportunity to trade rare pelts for European commodities. As anyone familiar with the North American fur trade knows, this is exactly what happened across most of this continent (for an interesting if far-fetched solution to this dilemma, see C. Martin 1978). Much recent research, in fact, indicates that Indian populations participating in the fur trade acted to maximize their own profits; the notion that indigenous cultural systems precluded the operation of the profit motive now seems quaintly romantic to many scholars (Tough 1990).

The most commonly cited example of hunting peoples living in harmony with nature is probably that of the North American Plains Indians. Yet the Plains buffalo economy that vanished in the late nineteenth century had originated only some 100 years earlier. During the late 1700s and early 1800s increasing numbers of tribes abandoned both Western deserts (the Plains Shoshone) and Midwestern woodlands (the Lakota) to pursue bison on the grasslands. No harmonious relationship between prey and predator could have been established in this short, demographically unstable period (Dasmann 1988:278). The Cheyenne, a quintessential buffalo tribe, did not even abandon farming to become full-time hunters until 1850. But with the establishment of trading posts on the high plains, the Cheyenne quickly learned how profitable buffalo hunting could be, and within years they were selling tens of thousands of hides at $3 each (Moore 1987:138–39). Contrary to popular imagination, they clearly could not have used every ounce of every bison carcass at this rate of slaughter.

Given time, the plains bison hunters might eventually have developed a sustainable economy. It does seem to be true, however, that bison had been unable to survive in the American Southeast until the seventeenth century. Before the introduction of European diseases, Indian populations had been too dense, and hunting pressures too severe, for the species to persist in this area. Only as human numbers plummeted were bison able to return for a short period (see Rostlund 1960).

Even when they do not overkill their prey, hunting peoples alter their environments profoundly. The majority of historically documented hunter-gatherers habitually burned their surroundings in order to clear out scrubby vegetation, to enhance the growth of certain desired plants, and, in general, to make hunting easier. The vegetation changes wrought by repeated burning are not necessarily destructive, but they certainly challenge the idealization of the pristine landscape. Moreover, they also show that preindustrial human beings cannot usefully be regarded as just another species. Fire may be a naturally occurring phenomenon in many landscapes, but human-induced burns are often systematic enough to cause massively unnatural transformations. Large areas of the American Midwest were apparently converted from heavy forest to tall-grass prairie under the continual pressure of Indian fire-setting; in Australia, the "firestick" alterations of the preexisting vegetation patterns were if anything more pronounced (Pyne 1990; on early burning in general, see Stewart 1956; H. T. Lewis 1982).

Such patterns of change in human-environmental relations rarely surface in the eco-radical literature, which shows only the vaguest conception of nature's historical geography. In a discourse of balance and harmony, history vanishes entirely; where all primal peoples are presumed
to have shared the same basic relationship with nature, geography disap-
pears as well. Eden is not only timeless—it is evidently placeless too.

Social Harmony among Hunter-Gatherers?

The assertion that hunter-gatherers are invariably peaceful and sexually
egalitarian is also unsupportable. Some of the most egalitarian of small-
scale societies have also been cursed with some of humanity's highest
rates of murder (Knauft 1987). Among the !Kung San—often upheld as
the paradigm of primal virtue—men often dominate women (Konner and
Shostak 1986:73), while murder rates are similar to those of most modern
industrial societies (Cohen 1989:92). In one central Australian hunter-
gatherer society, conditions have been considerably worse. As Mary
Douglas (1966:141) explains, "for the least complaint or neglect of duty,
Walbiri women are beaten or speared." Among the Eskimo even war was
not unknown, and if battles were small-scale affairs they could still be
quite bloody (Chance 1990:25). More striking is the incontrovertibly
dominant status of Eskimo men. Birket-Smith (1971:157) claims that
among the Netsilik tribes, "the killing of female children is so common
that a girl who is not betrothed at birth is usually doomed." The same
scholar's report on Netsilik adultery is equally telling: "when a man
punishes his wife for being unfaithful it is because she has trespassed
upon his rights; the next evening he will probably lend her himself"
(1971:158).

Many historical hunter-gatherers also habitually raided their sedentary neighbors. In pre-Columbian Meso-America, for example, the agrarian civilizations of the Basin of Mexico suffered repeated devastations at
the hands of the northern "chichimecs," a congeries of foraging peoples
described as fierce barbarians by anthropologist Richard Adams (1977:
269). In the American Southwest too, hunter-gatherers commonly plundered their sedentary neighbors, although the enmity between Pueblos
and Apaches was probably exaggerated by an earlier generation of scholars (Goodwin 1969). While hunter-gatherers are often peaceful among
themselves, this does not necessarily preclude them from exploiting
their less-mobile neighbors.

Nor were all hunter-gatherers affluent in the sense of enjoying abundant leisure and good health. This thesis rests largely on evidence from
the !Kung San of the Kalahari, a seasonally dry savannah that has been

erroneously called a desert. Hunters living in less-productive environments, such as the arctic tundra, present a grimmer picture. In fact, among virtually all documented hunting and foraging groups, as Mark Cohen (1989:130) demonstrates, "hunger has clearly been at least a seasonal problem . . . and starvation is not unknown" (see also Johnson and Earle 1987:33).

When examining life expectancy—surely an attribute of affluence by most accounts—the evidence is incontrovertible. While most hunter-gatherers historically suffered relatively little from malnutrition or infectious disease, the inescapable fact is that they have always died at remarkably young ages. Life expectancies for both historical and modern groups hover around twenty-five to thirty years (Cohen 1989). Even Blainey, a strong proponent of the primal prosperity thesis, admits as much: "The inability of most [Australian] aboriginal groups to care for many of their hapless members was one of the flaws in their standard of living. A relatively high death rate was probably necessary in order to guarantee that their material comforts persisted" (Blainey 1975:228).

Most fundamentally, the notion that contemporary hunter-gatherers are in any significant way primal is itself open to serious doubt. The !Kung San are not "living representatives of the Pleistocene" but rather a people who once practiced extensive pastoralism and were later deprived of their herds (Wilmsen 1983). Archaeological evidence now shows that the entire Kalahari was imbricated in global trading circuits hundreds of years ago (Wilmsen and Denbow 1990)—so much for the cinematographic myth of the coke bottle tossed down by crazy gods. The hunter-gatherers of the Southeast Asian rainforests, for their part, have been aptly described as professional primitives; their adaptation, it turns out, has depended for centuries on widespread trade in rainforest products. Without participation in commercial exchange, it is uncertain whether such peoples could ever have obtained adequate carbohydrates to sustain their populations (R. Fox 1969; Headland and Reid 1989). Significantly, as Southeast Asia's professional rainforest hunters have obtained modern weapons, marketable wildlife has not infrequently disappeared (see Candelina 1988).

Most glaring is the case of the Tasaday, famed for both their gentle ways and their total freedom from corrupting exterior contact. The discovery of this stone-age remnant now appears to have been an outright fraud. As Berreman (1991:34) concludes, "The evidence leaves no doubt in my mind that the entire Tasaday episode has been a deliberate deception, a hoax—not complex, but simple; not clever, but crude." Manipulated by a cynical Philippine government agency, "vulnerable villagers . . . were induced to cavort, clad in leaves, as cave-dwellers before outsiders during brief, preannounced visits" (Berreman 1991:34). That the scholarly community so misconstrued this tiny group of people testifies rather dramatically to the power of our common longing to find a living Eden.

As the preceding discussion shows, the term "primal" proves just as insulting as its predecessor, "primitive." Despite the wishful thinking of eco-radicals, we must recognize that all human societies have developed along their own pathways, and that no one group is any more primal than any other. History is an attribute of all human societies. In recognition of the hazards of labeling, many scholars now prefer the purportedly neutral term "indigenous," but this becomes extremely problematic once we begin to examine the history of human migration. The least offensive and most descriptive term, I would propose, is simply "small-scale society."

But regardless of the terms employed, small-scale, nonindustrial societies remain powerfully attractive to those who wish to recover a lost paradise. As demonstrated above, such an idealization is no longer tenable in regard to hunting-gathering societies. A sizable majority of documented small-scale societies, however, have long supported themselves primarily from cultivation. In order to counter the anti-humanist anarchist primal harmony thesis, which is more forgiving of farming than that of the primitivists, it is necessary to examine the environmental and social records of purportedly harmonious horticultural and agricultural peoples.

In order to make sense of the vast array of subsistence patterns found among nonindustrial farming societies, a quasi-evolutionary schema will be employed in the following discussion. It thus begins with the origin of agriculture, proceeds to analyze groups practicing extensive forms of cultivation, moves on to consider intensive agriculture in tribal societies, and concludes with a brief discussion of peasants in the preindustrial state. Particular attention will be placed on the issue of population growth, which seems to be one of the key forces behind environmental degradation, as well as one of the prime movers behind social change.

The Transition to Cultivation
Even if hunting and gathering is not a sure path to a leisure-filled life, most anthropologists agree that it is generally less demanding than agriculture. The domestication of plants and animals is now more often seen as a necessary but painful expedient than as the "great discovery" trum-

peted in the past. While theories of agricultural origins vary greatly, one of the most powerful holds population pressure and the corresponding depletion of resources as key initiators (see Cohen 1977). As local hunter-gatherer populations expanded they began to exhaust local sources of wild foods. Eventually, if they were to survive in the same habitat, such a stressed group would begin to manage key species in order to intensify production, leading ultimately to domestication. The traumas associated with the shift to agriculture may have been quite severe; Mark Cohen (1989:127) concludes that human life expectancies reached their nadir, at a shocking eighteen years, precisely at this juncture.

The patterns of human-environmental harmony and disruption found among tribal peoples are not random. One could cite a long list of cultural attributes that contribute to either the development of a conservation ethos or a widespread disregard for nature's preservation. But population density by itself seems to be an overwhelming factor in many instances. When human numbers in a local area grow too large for a given mode of subsistence to accommodate, serious degradation often follows.

Except in a few blessed environments, hunting and gathering is possible only so long as the human presence remains extremely sparse, usually less than one person per square mile. Since many hunter-gatherer populations do increase over time, albeit very slowly, such density thresholds seem to have been reached many millennia ago in several parts of the world. As this occurred, hunting and gathering modes of life became untenable, and local populations either crashed or perforce adopted some form of cultivation or pastoralism.

Tribal Cultivation and Environmental Degradation
Tribal horticulturalists, at least those living in sparsely settled wooded environments, typically practice swidden, or slash-and-burn cultivation. In the typical swidden system, each family cuts and burns a section of woodland every year, relying on the resulting ash to form a fertile seedbed. After a few years, as fertility declines and pests and pathogens multiply, they abandon the plot to natural succession, leading eventually to reforestation. In a properly functioning swidden system, once forest vegetation has been fully reestablished the plot may again be burned and the cycle run anew.

While anthropologists and geographers in the 1960s and 1970s often pictured swidden as perfectly fitted to the forest environment, it is now abundantly clear that this is true only under certain conditions. The key to successful swiddening is simply a relatively low population density. If local human numbers grow too large, fallow periods must shorten; if population growth is relentless, rest periods will eventually become so brief that forest regeneration ceases to occur. If the process continues unabated, incessant burning may result in progressive soil deterioration and thus to the long-term degradation of plant communities. Indeed, many of the botanically impoverished grasslands found scattered through regions otherwise supporting tropical rainforests are now believed by many to be the consequence of maladaptive swidden cultivation (Seavoy 1975; Scott 1977). Local societies are sometimes cognizant of the process yet unconcerned. As Seavoy (1975:50) bluntly writes, "The agriculturalists of Kalimantan have little reason to care if a whole forest is destroyed as long as unoccupied forests are nearby."

As the pendulum of scholarly fashion begins to swing back from romanticism, swidden cultivation is losing the warm glow painted around it by scholars during the 1960s and 1970s. The notion that slash-and-burn cultivation "mimics the tropical rainforest" has been discredited; most Amazonian swiddens, for example, are composed not of multistoried, polycultural crop melanges, but rather of concentric rings planted to single crop species (Beckerman 1983). Moreover, the low labor requirements of swiddens in the Amazon and New Guinea, it turns out, have been made possible only because of trade with the world economy. In the indigenous systems, trees had been felled by stone axes, a tremendously laborious process; only with steel blades does the operation become relatively easy.

Nor is swidden as energy efficient as it was once believed to be. While studies in the 1970s showed that swidden cultivators could extract an admirable ten to twenty calories of food energy for every calorie expended in working the fields, it was later realized that such figures completely ignored the fantastic quantities of energy that go up in smoke every time a plot is burned, an "oversight comparable to omitting the gasoline used . . . [in] American farming" (Rambo 1984:156). Revised calculations suggest that an extremely meager return of .005 calories of food per calorie expended is more accurate (Simmons 1989:131). Moreover, the burning of swidden fields is an inherently polluting process. As Terry Rambo (1985) now argues, small-scale societies not uncommonly contaminate their own environments by swidden burning and by other practices as well. He writes: "Pollution of the air by domestic fires, pollution of water by introduction of disease organisms or fish poisons, pollution of soil by indiscriminate dumping of rubbish, are all environmental impacts which are qualitatively no different than those so loudly deplored when they result from the activities of civilized societies" (Rambo 1985:78).

Finally, several studies indicate that many swiddeners lead arduous rather than idyllic lives. Reid (1987:39–40), for example, suggests that certain Southeast Asian "hill tribes" historically had markedly low birth rates in part because the women had to work so hard that they impaired their own fertility. In northern Luzon, contemporary Kalanguya-speaking swidden cultivators described their work to me as extremely taxing, and all eagerly awaited the coming of roads so they could take up commercial vegetable farming (M. Lewis 1992).

Social Relations among Tribal Cultivators

Although the literature of deep ecology would have one believe that all swiddening peoples are peaceful and egalitarian, the social relations found among them actually exhibit tremendous variation. Some tribes, such as the Buid of the Philippines (Gibson 1986), are extraordinarily peaceful; others, like the Yanomamo of Venezuela, are among the most violent societies the earth has known (Chagnon 1968). Owing to Yanomamo fame within anthropology as the earth's quintessential "fierce" society, Yanomamo men's proclivity for mayhem requires additional scrutiny if we are to come to terms with the myth of primal social harmony.

Anthropologists have debated the cause of Yanomamo violence with some heat. Marvin Harris (1974) argues that the scarcity of protein in the form of wild game leads to fierce competition among Yanomamo men. The difficulty of obtaining adequate protein in Amazonian diets is, however, a controversial issue (see Gross 1975; Beckerman 1979; Chagnon and Hames 1979). Napoleon Chagnon (1990), on the other hand, the most noted ethnographer to work among the Yanomamo, believes that the principal impetus for their habitual violence stems from a shortage of women combined with the sociobiologically based imperative to procreate. The majority of anthropologists, however, firmly deny the validity of such sociobiological reasoning. Bruce Albert (1989) has even accused Chagnon of exaggerating Yanomamo violence due to an unscientific Hobbesian prejudice against tribal groups, whom Albert accuses Chagnon of construing as "our contemporary ancestors." But this critique hardly furthers the "primal cause." Albert (1989) denies that Yanomamo violence is exceptional in part by presenting evidence that other Amazonian peoples inflict similarly high mortality rates on each other in their own skirmishes.

Whatever the root cause, it is clear that the world of the Yanomamo is anything but peaceful. Perhaps, as Harris argues, the Yanomamo men may strike a certain balance with nature by killing each other off, but I doubt whether their incessant raids and their head- and chest-bashing rituals—not to mention their beating and raping of women—would be appealing to deep ecologists and eco-feminists.

While the Yanomamo are among the most consistently brutal of all peoples, neither male exploitation of women nor unconstrained hostility toward outsiders are at all uncommon in the anthropological records on tribal cultivators. The oppression of women may reach an extreme among certain groups of Northeast Africa, in which all girls are subjected to horrific genital mutilation (Lightfoot-Klein 1990). And when examining intertribal relations, one would do well to consider the Tugarians of New Guinea: "Like their Viking counterparts, these pirates used long, fleet canoes, stealing along the coast and up the creeks, plundering, murdering, and taking prisoners. In order to prevent the prisoners from fighting or running away, the Tugarian warriors would break their arms and legs, keeping them alive, however, so their supply of meat would stay fresh until it was needed" (Sagan 1974:3).

Like most lowland Melanesia peoples, the Tugarians are relatively egalitarian within their own villages. Yet such egalitarianism in virtually all of these societies "stops short of relations between the men and women" (Strathern 1987:2). While Melanesianists have shown that women typically have their own spheres of power and influence, the overall domination by men is unmistakable. As Keesing (1987:35) writes of the Kwaio of the Solomon Islands, "Women are consigned the tasks of sustained labor in swiddens, raising pigs, cutting firewood and carrying water, childcare and cooking. Men's lives are centered on ritual, feasting and exchange, and the public politics of litigation."

Some tribal agriculturalists, it must be noted, come close to achieving gender equality. But as long as a single society in a given region follows a path of violence, any gentle, irenic, nonsexist peoples in the vicinity may find themselves being viciously exploited anyway. A case in point are the Buid of Mindoro Island in the Philippines. The Buid seldom harmed anyone (Gibson 1986), but they found themselves continually harassed by martial, slave-raiding groups from the southern islands. Over the centuries vast numbers of tribal horticulturalists from all accessible areas of the Philippines were carted off by these pirates to be sold as chattel in the the Sulu Archipelago (J. Warren 1985). Because of such terror, many peaceful Philippine peoples were forced to flee to remote mountains where they were hard pressed simply to survive (J. Warren 1985:169).

European prehistory should also be reexamined in the same light. Perhaps, as Gimbutas argues, the old Europeans were remarkably peaceful and egalitarian. But, according to her own scenario, when male-dominated, militant, hierarchically organized, proto-Indo-European-speaking pastoralists swept out of the Ukrainian steppes, they were utterly overwhelmed.[2] Only on isolated islands, like Crete, did the old culture persist, and even here it would be vanquished as soon as Greek-speaking Dorian pirates took to the seas. As long as any one group follows a path of violence and exploitation, peaceful societies will be easily victimized and not uncommonly completely annihilated.

As the preceding discussion shows, small-scale societies practicing swidden and other forms of extensive cultivation do not necessarily exist in the social and ecological idyll imagined for them by deep ecologists. It remains true, however, that most of these groups cause *relatively* little environmental degradation. Yet such benign interactions with nature are often a function more of their low population densities rather than of their special affinity with the natural world. As population density increases, tribal cultivation usually begins to extract a more substantial environmental toll.

Intensive Agriculture in Small-Scale Societies

Many small-scale societies, it must be said, do maintain low population densities over long periods, thereby presenting modest environmental threats. Others grow steadily but avoid degrading their farmlands by devising more intensive cultivation systems that obviate the need for long-term fallow (Boserup 1965). Permanent cropping, however, almost invariably demands fertilization, hence heavy applications of labor. Whatever primal affluence certain swidden cultivators may have enjoyed vanishes once their populations grow too large for forest farming to support.

But even if they have successfully intensified production, tribal and peasant societies characterized by permanent-field agriculture detract from nature in other ways. True, many do devise sophisticated farming techniques that maintain high productivity levels over many generations. But often a creeping level of deterioration is still evident. As nutrients are slowly extracted from forests and meadows, in the form of ash, leaf mulch, and "green manure," in order to subsidize fields, uncropped areas may be gradually impoverished. The stunted forestlands of Korea and much of southern China, for example, may owe their degraded state in part to an imperceptibly slow leakage of nutrients to the cultivated fields. As Peter Perdue (1987:35, 247) shows, peasants in south central China burned entire forests in order to sell the fertilizing ash in downstream areas, a process that contributed directly to what he aptly calls the "exhausting" of the earth.

Tribal farmers who practice metallurgy—often necessary for labor efficiency in cultivation—also degrade their forests, and foul their atmospheres as well, by operating smelters. Goucher (1988), for example, cites evidence for ancient, widespread deforestation in a region of Togo noted for its ironworks. She describes the corresponding air pollution in clear terms: "Nineteenth-century German travelers in the Bassar region noted that the concentrated, low-lying smoke and fumes of smelting furnaces were visible at a considerable distance. Evidence for the environmental effects of heavy particulates, oxides, fumes, ash, dust, and gaseous pollutants is now being studied in changes in soil profiles recording the last five centuries of industrial activity" (1988:60).

Wherever human population growth continues, new threats continue to emerge. Farming innovations, most of which require ever greater applications of labor, are necessary if adequate food is to be obtained without destroying the land's productive capacity (Boserup 1965). Yet even if soil fertility is maintained, wildlife invariably diminishes. As populations mount, ever larger tracts of land must be devoted to provisioning the human community. In such circumstances wild animals become competitors for scarce resources — as well as occasional sources of rare protein.

Owing to their labor burdens, the lives of tribal peoples practicing intensive cropping are usually quite arduous. Not surprisingly, individuals in such societies often suffer from poor health and are weakened by malnutrition and protein deficiency. Sanitation is often miserable and gastrointestinal parasites rife, while skin and respiratory diseases commonly make life miserable. Dennett and Connell (1988) have convincingly shown how a diet based largely on sweet potatoes formerly resulted in atrocious health standards for a group of New Guinea highlanders. As they argue, these people "have no wish to retreat to the 'subsistence affluence' and nobility that have sometimes been thrust upon their ancestors" (1988:281).

High-density, small-scale societies, contrary to their depictions in the literature of the deep ecologists, are also not uncommonly stratified to a high degree. Among the Ifugao of the Philippines, for example, wealthy rice-terrace owners wield great powers over the landless poor; earlier in this century they often sold their debtors off as slaves (M. Lewis 1992). In many parts of Melanesia, where status is generally earned rather than inherited, ambitious "big men" often denigrate the less successful as "rubbish men." The "big man complex" usually develops in Melanesian societies marked by high population density, competitive exchanges, and the accumulation of wealth, but where such conditions and institutions are lacking, "great men" typically emerge by cultivating power in the ritual sphere. Interestingly, while the latter societies are more egalitarian than the former, they are also typically more patriarchal (Strathern 1991).

Peasants in the Preindustrial State

Although neither increasing population density nor internal social evolution leads automatically to the emergence of large-scale social forms, it remains a historical fact that over time progressively larger areas of the earth have been incorporated into state-level formations (in other words, they have been detribalized). Since some eco-radicals believe that even in nonindustrial states we can find evidence of profound ecological harmony, we must turn now to consider the environmental relations characteristic of peasant societies.

The environmental records of preindustrial states founded on peasant cultivation are generally dismal, especially in the realm of wildlife conservation. Writers who laud European peasants for having lived harmoniously on the land should acquaint themselves with the history of European wildlife management. Most large mammals that survived in Europe did so only because the aristocracy so enjoyed slaughtering them that they maintained hunting parks and established game regulations. Under Norman law, for instance, a commoner could lose a year's liberty merely for causing "a stag to pant" (Graham 1947:17). Had the peasants (and their poacher heroes like Robin Hood) had their way, such troublesome and succulent creatures as red deer and wild hogs likely would have vanished from the British Isles centuries ago. This is not to credit feudal lords for their environmental wisdom; rather it is to demonstrate the inescapable ironies that confront anyone who would hold up peasant societies as ecological exemplars.

Those who dismiss such European examples on the grounds that Western societies have a special talent for rapine need to look no further than the history of wildlife extermination, deforestation, and soil erosion in China. At the dawn of Chinese civilization, much of the North China Plain and the Yangtze Valley was home to tigers, rhinoceroses, elephants, and many other large mammals. Such creatures were locally exterminated by relentless hunting pressure, in this case a crime perpetuated in large part by the aristocracy. While some Chinese philosophers bemoaned the loss, it is notable that such protests came as much (if not more) from the rationalistic and secular Confucians (Gernet 1968: 74) as from the Taoist, beloved of the deep ecologists for their naturalistic creed. In fact, ever since the early Chou, China proper has been a faunal wasteland. And the destruction of what little Chinese wildlife as remains continues to this day, well-publicized preservation efforts such as the panda campaign notwithstanding (see Smil 1984; on discrepancies between Chinese attitudes and behaviors, see Tuan 1974).

East and West, the story is similar wherever one looks. Large animals do not survive in any appreciable numbers in areas of dense human habitation. Tigers, which once ranged from the shores of the Caspian Sea east to the island of Bali and north into the Siberian forest fastness, survive today only in a few scattered sanctuaries. Even in India, which deserves great praise for its struggle to preserve the species, tiger populations have only been maintained by removing entire villages from tiger habitat—a policy bitterly denounced by socialist environmental advocates (Guha 1989). The antihunting sentiment found in most Hindu castes (although notably not among the lordly Rajputs) has undoubtedly resulted in greater wildlife survival in India than could be expected on the basis of population density alone. But even here one is confronted with an inescapable conflict between the needs of human beings living in a high-density social environment and the survival of other large mammals.

In short, few peasant populations leave room for anything natural that does not directly benefit the human community. One can hardly blame peasants for this; they have been, after all, so widely victimized by more powerful social classes that their lives have typically verged on desperation. But to visualize peasants as ecological exemplars is as much a delusion as imagining medieval peasant life as fat, merry, and free (see below, chapter four).

Population Growth and Ecological Degradation

The challenge I have presented to the thesis of primal harmony emphasizes the role of population growth in disrupting whatever balance may exist locally between human communities and the natural world. Most radical environmentalists also point to population pressure as one of the gravest threats faced by nature. Yet many seem to believe that this phenomenon is unique to modern times. In fact, much evidence points to the end of the upper paleolithic as the beginning of the human population advance (Simmons 1989:83).

Admittedly, before the industrial revolution, human numbers grew slowly and fitfully. Some areas saw long stretches of virtual stasis; Egypt, for example, fluctuated between 3 and 5 million persons for several millennia (McEvedy and Jones 1978:227), while Africa's population stagnated for hundreds of years, due mainly to the depredations of the slave trade. Other regions, notably North and South American and Oceania, saw their populations plummet after Eurasian diseases appeared in the post-Colombian period (Crosby 1986). But despite such intermittent downturns, the global pattern has been one of an ever rising human tide starting before the neolithic revolution. In the year 8000 B.C.E. the earth probably supported some 5 million human beings; by 400 B.C.E. this number had increased to 150 million, and by C.E. 1800, in the earliest phase of the industrial revolution, it stood at 720 million. Slowly but steadily, ever less room and ever fewer resources were available for non-human species. Had the industrial revolution never occurred, this process may well have eventually destroyed remaining wildlife habitat just the same.

Those who believe that human economic and technological progress is worthwhile must recognize a beneficial side to population growth—*to a point*. Relatively dense populations in certain areas allow the economies of scale and the potentialities for specialization upon which civilization rests (Boserup 1965). But even those who believe with the primitivists that civilization has been a hideous mistake must still account for the inescapable human propensity to procreate. Even if 99.9 percent of the earth's human population could be wished away so that the remainder could return to hunting and gathering, history suggests that some local group would eventually outstrip its food supply and thus set the process into motion once again.

Yet it remains true that the magnitude of the current population problem is unique to modern times. Populations did explode in Europe in roughly 1800 and in most of the rest of the world by 1900. Yet many European countries are now beginning to experience demographic decline. Radical environmentalists are absolutely correct in arguing that the human burden on the earth is already too large. But the key to slowing down and eventually reversing the population tide lies not in returning to nature but rather in industrializing fully.

The preceding discussion has traced out a historical, developmental trajectory, beginning with Pleistocene hunters and ending with peasants in the preindustrial state. To conclude this chapter, I will return briefly to the matter of tribal society in order to provide a more detailed account of human-environmental relations in one particular modern-day, small-scale group. Through such an extended analysis the notions of the primal mind and the tribal way of life so favored by eco-radical rhetoricians may be more thoroughly discredited. But the reason for this gambit is also partly personal, for my own views of tribal life have been indelibly inscribed by extended fieldwork among the Southern Kankana-ey, a horticultural people inhabiting the mountains of northern Luzon in the Philippines. While I have come to have great respect for the Kankana-ey as a people, living among them has made it impossible to hide from the fact that they have despoiled their environment to a remarkable degree (for an extended discussion, see M. Lewis 1992).

Environmental Degradation in Northern Luzon
The inhabitants of the rugged mountains of northern Luzon are classified as tribal by most observers. They were never fully subdued by the Spanish colonialists, and they retain a large array of indigenous customs to this day. Their traditional political unit, the autonomous village community, remains the site of considerable local authority. Village elders generally maintain control of land tenure, and, in many areas, all other legal proceedings as well. Through most of Luzon's northern highlands, agriculture is still subsistence oriented, based on harvests obtained from small swidden fields and tiny irrigated rice-terraces carved out of steep mountainsides. Participation in the market is secondary to domestic production, except in the highland's southernmost reaches. Old ways of life are, however, threatened in many villages by the forces of modernity. In the less-developed central and northern regions many individuals have taken up arms to demand political autonomy in a desperate attempt to retain control of their own resources and way of life.

Located in the southern mountains, the village of Buguias, site of my field research, is commercially oriented and politically complacent. But it is also noted throughout the region for its retention of indigenous cultural forms. Most Buguias villagers reject Christianity, proudly referring to themselves as Pagans (or *"Paganos"*). Buguias Paganism revolves around ancestor worship, which in turn is made manifest in innumerable and often massive redistributive prestige feasts. In these ceremonies, celebrating couples invite neighbors, relatives, and, most importantly, deceased ancestors, to receive food and alcohol and to make visible their communal links through sacred dance.

The Buguias region also has a well-deserved reputation for peacefulness. Early in the century, when many of their neighbors were fervent headhunters, the people of Buguias were widely considered shy and retiring. Their aptitude for peace is reflected in a traditional jury system that operates at full strength to this day. Virtually all local conflicts are settled amicably, with the community's body of elders consistently devising settlements that mollify the contesting parties. Indeed, compromise and conciliation are deeply embedded cultural ideals; on this score, the people of Buguias have much to teach American society. Merely to engage a lawyer in Buguias is to break a fundamental social covenant.

But like any community, Buguias has a negative side. This is especially clear in the realms of class and gender. While women can have considerable economic clout, religious, political, and jural power is almost entirely in male hands. Nor is power evenly distributed; since the earliest remembered times a handful of rich men have dominated the entire community. In earlier generations, local plutocrats even kept slaves, usually purchased from the Ifugao people (the world's most famous rice-terrace builders) living to the east.

Even Buguias's peaceful inclination seems to be of relatively recent origin. While most modern residents deny that their ancestors ever hunted heads, Spanish records as well as one of their own rituals indicate that they did. The community's irenic bent seems to have developed only after its economy was reoriented some six or seven generations ago toward interregional trade. Once this occurred, social evolution in Buguias took a pathway unusual for a small-scale society.

Ever since the mid- to late 1800s, when they assumed the role of trade intermediaries for a large swath of the southern mountains, the Buguias people have had a mania for capitalism—albeit capitalism as interpreted through their own cultural system. Indeed, the whole point of their elaborate and remarkably expensive redistributive rituals is to enlist supernatural aid for the further accumulation of capital; feasting one's neighbors is actually something of a by-product. It is not without reason that members of the community's Christian minority bitterly accuse Pagans of worshipping money. Indeed, in several significant rituals, old silver peso coins, as well as blood-soaked taro chips symbolizing cash, command considerable attention.

Environmentally, Buguias is a disaster area. Before the Second World War, cultivation had been subsistence oriented and little overt land degradation was evident. But even if cropping had been sustainable, wildlife had either disappeared or become very rare due to over-hunting and population pressure. Monkeys, for example, were exterminated throughout the region generations ago. The local folk are not proud of this; one touching story tells of a hunter weeping after witnessing the bereavement of a monkey whose "spouse" he had just slaughtered. Yet unsustainable hunting continued and wildlife progressively disappeared. Today there is very little left; even songbirds, still avidly pursued, are scarce. Many elders would like to institute conservation measures, but the young men continue to kill virtually anything that moves in order to eat rich morsels when they drink their cheap gin.

After the Second World War, the Buguias villagers wholeheartedly adopted commercial vegetable farming, with no outside prodding. Local entrepreneurs realized that good profits could be made and they wasted no time learning the new techniques. Once they found success, their neighbors followed suit, abandoning subsistence crops to plant cabbage, potatoes, carrots, lettuce, and other temperate vegetables that they could sell at high prices in lowland markets.

Successful truck gardening, it soon becomes apparent, required the massive use of a wide spectrum of insecticides, fungicides, and herbicides. Many streams are now so polluted with these poisons that they no longer support vertebrate life. Recently the community's richest men discovered fat profits in hiring bulldozers to flatten large fields in the higher-elevation cloud forests. Seldom do they even bother to salvage the topsoil; if enough chicken manure (imported from the lowlands) and synthetic fertilizers are applied, good harvests can be obtained from the easily worked subsoil. As bulldozing proceeds, erosion accelerates and water tables drop. Yet most residents continue to regard the wealthy men responsible for this devastation as the *progreso* benefactors of the larger community.

Despite both environmental destruction (the extent of which many residents recognize) and the recent stagnation of both the vegetable industry and the larger Philippine economy, the entire Buguias community fully endorses the postwar transition from subsistence cultivation to market gardening. The reason is simple; people are much more prosperous now than they were under the old regime. Several elders, from whom I had hoped to learn aspects of local history, would not even discuss the old days. One woman simply retorted that "life was terrible—we only ate sweet potatoes."

My time in northern Luzon demolished my most cherished beliefs. Not only had I previously espoused all of the tenets of radical environmentalism, but I had been fully inculcated in a school of cultural geography that regards indigenous peoples as lacking commercial sensibilities and as always respectful of their environments. In subsequent writings on the subject (1989; 1992), I have struggled with one daunting question: had I simply happened upon an aberrant exception to otherwise valid rules or did these findings mean that our conventional notions about small-scale socioeconomic organization, notions most strongly articulated in the eco-radical literature, require substantial revision? Ultimately, I concluded that the best answer is simply that the typical small-scale society is a mirage. Far more variability marks the realm of tribal social organization than eco-romantics would dare recognize.

Conclusion

A large proportion of eco-radicals fervently believe that human social and ecological problems could be solved if only we would return to a primal way of life. Ultimately, this proves to be an article of faith that receives little support from the historical and anthropological records. Although many radical environmentalists are anxious to find empirical groundings for their primal visions, their marshaling of evidence is far too selective to satisfy the demands of scholarship. Meanwhile, in academia the tide has finally turned. The contemporary view of careful scholars is well summarized by Timothy Silver, who concludes that American Indians on the whole were neither despoilers nor preservers of nature, and that "since his arrival in North America, mankind has remained apart from, and altered, the natural world" (1990:66, 197).

To counter the thesis of primal purity, I have attempted to construct a more systematic survey portraying a wide array of small-scale social forms. Of course, I too have carefully selected evidence fitting my larger argument, conveniently ignoring conflicting accounts. But unlike the deep ecologists, I am not trying to construct a model of the essential human condition encountered in prelapsarian societies. Based on a conviction that the human condition—in all forms of socioeconomic organization—is characterized by diversity, the argument presented above is not vulnerable to contradictory examples. If I have presented a rather tiresome parade of human iniquities it is to show only that the human destruction of nature and exploitation of fellow humans are facts of *long* standing, not that small-scale societies are *necessarily* destructive or exploitative.

If we are to construct an environmental movement powerful enough to enact needed reforms we must first relinquish our romantic fantasies. A meaningful environmentalism cannot be based on nostalgia, wishful thinking, and faith that the inherent goodness of humanity will manifest itself once civilization is dismantled. As the following chapter will demonstrate, romantic nostalgia is also strongly evident in the eco-radical notion that scale is paramount in determining whether ecological sustainability can be realized.

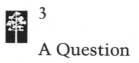

3

A Question

of Scale

■ The Radical Position

Small Is Beautiful

Radical environmentalists have argued since the late nineteenth century
that large-scale economic and political structures are both inherently
dehumanizing and deadly to nature (Bramwell 1989). In the 1970s the
idea that all organizations should be small of scale was eloquently re-
stated by the economist E. F. Schumacher, whose *Small Is Beautiful*
(1973) remains an environmental classic. Schumacher and his followers
believe that expansive social entities are invariably governed by stifling
bureaucracies whose rule-bound behaviors lead to environmental degradation and social waste. True human values, they aver, can only be
realized in intimate groups. Schumacherians have also argued that the
wisdom of small-scale organization is mirrored in ecological systems,
themselves structured around local transfers of energy and matter.

In recent years Schumacher's vision has been most effectively forwarded by Herman Daly, recently appointed senior economist in the
environment department of the World Bank (an organization that evidently believes in hiring its critics). Daly, along with theologian John
Cobb, argues that scale is paramount in both human affairs and human-
environmental systems (Daly and Cobb 1989). Only by reorienting our
society and economy along local lines, they claim, can we maintain
healthy community relations and avoid the incessant growth of productive capacity that will eventually overwhelm the planet.

The glorification of the small and the vilification of the large have wide currency in eco-radical circles. Such notions are implicit in primi-tivist ideology, since hunter-gatherer societies are by necessity small and localized. Antihumanist anarchists essentially agree; according to Ros-zak, the common enemy of "person and planet" is simply "the bigness of things" (1979:32–33). Humanist eco-anarchists argue along a similar line, with Bookchin (1989:185) urging us to "rescale communities to fit the natural carrying capacity of the regions in which they are located" (1989:185). Radical eco-feminists similarly stress the need to reestablish intimate communal relations and to shun large-scale bureaucratic struc-tures. The most pointed attack on complex, centralized institutions, however, is probably that of Jeremy Rifkin, legal gadfly of the eco-radical movement. Rifkin regards all large-scale structures as inescapably self-destructive, charging that they will eventually require more energy for their maintenance "than the system can afford" (1989:107). Only eco-marxists, it would seem, diverge from the standard eco-radical position that beauty comes *only* in small packages.

Political Devolution
The yearning for small, decentralized social entities has profound conse-quences for radical-environmental political theory. In an ecologically enlightened future, sovereign states must devolve into polities small enough to be fully guided by participatory democracy, giving every cit-izen equal and immediate say in all political affairs. Such truly demo-cratic and autonomous communities might join together to form vague federations, but actual power must remain fixed at the local level. Many eco-radicals view political decomposition as the only real alternative to the opposite course, on which we are now supposedly affixed, of ever mounting centralization. Our current path, Bookchin warns us, will culminate assuredly in an Orwellian nightmare of police-state brutality: "The fervent belief that liberty would triumph over tyranny is belied by the growing centralization of states everywhere and by the disempower-ment of people by bureaucracies, police forces and sophisticated sur-veillance techniques—in our 'democracies' no less than in visibly total-itarian countries" (1989:20).

Interestingly, the eco-anarchist view on this score is virtually identical to that held by the members of the marginal right-wing camp of so-called free market environmentalism. As Cuzan (1983:28) argues in regard to the public control of water resources: "The long run trend of public policies to expropriate water rights and centralize control over the re-source in federal and state bureaucracies can be explained by two natural laws of politics: the iron law of political redistribution, and the iron law of hierarchical centralization." Although conventional eco-radicals support politically mandated economic redistribution, their attitude toward governmental centralization is identical to that of individuals who (supposedly) occupy the opposite extreme of the political spectrum.

Scale in Economic Endeavors

In a down-scaled eco-radical world, economic organization would be drastically reordered. Autarky is the ideal, with each local region producing its own necessities. Interregional trade, except perhaps in a few scarce natural products like salt, should wither away to the greatest extent possible.

Eco-radicals disparage exchange both because they value intimate social connections untarnished by personal gain and because they believe that moving goods over long distances demands too much energy. Firms whose operations span the continents are especially reviled. Such organizations not only embody large-scale, depersonalized, bureaucratic social relations, but they also undermine local communities by subsuming them within their vast webs of profit and exploitation. Among all schools of radical environmentalism, one has simply to append "multinational" to "corporation" in order to move it from the category of the merely bad to that of the truly wicked.

Eco-radicalism further disparages transregional economic links because of their supposed powers of cultural demolition. When peoples of non-Western societies are exposed to the products of the industrialized world, such reasoning goes, they are soon seduced by the glamor of commodity consumption. Eventually they may abandon their own cultures and internalize the bankrupt values of the West. Especially villainous is the export of films, music recordings, and soft drinks. Along with such dross come advertisements—blandishments for cultural suicide. If present trends continue, eco-radicals fear, global culture will be massively homogenized, resulting in a dreary world filled with persons shallow of spirit, covetous of base consumer objects, and brainwashed by the latest Hollywood fantasies. As noted marxist geographer Richard Peet (1989:191) argues, "Entertainment is the opium of the people. Culture becomes the aesthetics of anaesthesia." Few eco-radicals would disagree.

These beliefs about scale are also implicated in the green radicals' opposition to advanced technology. Many new technologies, they argue, entail an ever greater centralization of authority, leading to a steady diminution of personal freedoms. In contrast, they favor "appropriate technology" as supporting decentralized structures of social power. Certain forms of solar energy, for example, allow individual residences to disconnect themselves from the public power grid, the veritable emblem of centralized authority's sinister web.

Several heretical enthusiasts have touted microcomputer networks linked by telecommunication lines as potential vehicles for participatory democracy (Kassiola 1990:208) or even grass-roots subversion, but most of the movement's members scoff at such suggestions as exemplifying technophilic naivete. Langdon Winner (1986), for instance, argues that computers only enhance the power of large firms and bureaucracies, allowing them to devise more sophisticated forms of surveillance and police control. Moreover, he continues, advances in telecommunications will encourage multinational corporations to transfer their operations to whatever areas of the world offer the lowest wages and most compliant workforces. As this happens, footloose firms will abandon whatever residual responsibility they may feel for local communities, resulting in ever greater human alienation and social isolation.

Globalism

Despite their love of all things small and local, many eco-radicals simultaneously espouse global political action (for example, Devall and Sessions 1985:37). The most pressing ecological problems, they recognize, are planetary in scope, requiring international coordination of some sort. Bramwell argues that this tension between radical environmentalism's local and global concerns reveals a deeply rooted paradox, one similarly evident in the fact that in earlier years "both scientists working for a planned world state *and* atavistic poets [came] together in the ecological movement" (1989:230). While the contemporary mood (exclusive of eco-marxism) shuns both planning and the state, the hope endures that some kind of global congress will spontaneously emerge from properly (un)structured local communities and regional confederacies.

Modern-day radical greens do not view their dual concern with small- and large-scale political frameworks as problematic. Rather, they accord each scale its own sphere of thought and action. Local problems demand the greatest concentration of effort, but if all eco-activist cells, united by common ideological and spiritual bonds, can cooperate, global problems too might prove manageable. Indeed, a reorganization of society at the local level, many believe, will itself greatly reduce planetary environmental pressures. This stance is captured perfectly in the slogan "think globally; act locally."[1] Here again we encounter eco-radicalism's philosophical idealism, its insistence that right thinking alone holds vast potential power.

Bioregionalism

As an extrapolation of these beliefs about the importance of organizing on a small scale, deep ecologists believe that healthy human communities must be explicitly tied to the natural world through the principle of "bioregionalism." The bioregional view is predicated on the notion that the earth may be divided into discrete ecological regions unified by the correspondence of physical features and life forms. If human societies were to organize themselves in accordance with these regional patterns, the argument goes, they would more readily find ecological harmony. The unit accorded primacy varies in different bioregional schemes, ranging from individual watersheds to physiographic provinces (the Ozark Mountains) to entire biomes (the Eastern deciduous forests). The foremost proponent of the philosophy, Kirkpatrick Sale (1985), explicitly arranges variably defined and differentially sized bioregional entities in a hierarchical order, with the "ecoregion" (a biome defined by "climax" vegetation) encompassing the "georegion" (the watershed), which in turn supersedes the more localized "morphoregion."[2]

The precise definition of the bioregion is an unimportant issue to the committed eco-radical. What matters is rather that autonomous human communities organize themselves around *some* small-scale, naturally constituted, geographical area. Only in such an intimate place-bound setting can people begin to understand their underlying relations with the natural world (Sale 1985:53). In turn, somewhat larger natural regions are expected to give shape to political confederacies emerging spontaneously through the peaceful interactions of autonomous villages. Of course, actual power would remain diffuse, with "nothing done at a level higher than necessary" (Sale 1985:94).

The bioregional principle, many eco-radicals affirm, has long been implicitly employed by primal peoples in devising their own sustainable modes of social organization. Sale (1985:60), for example, argues that the "tribal conglomeration of Algonkian-speaking peoples" corresponded perfectly to the ecoregion of the Northeast hardwoods. In fact, until the modern era began, such associations between human and natural geographical patterns were allegedly ubiquitous. According to Young (1990: 136), even some of the "pre-nation-state kingdoms of Europe" were coincident with bioregions. The task of the radical green movement thus once more becomes one of reimagining and then recreating a lost socio-ecological form.

Such a reinvention, many contend, is already under way. In the far northern reaches of California, for example, a multitude of "neo-tribes" constituted around individual watersheds are now coming together to

form the emergent "Shasta Nation" (see Devall and Sessions 1985:23).
From this and other such nuclei, proponents hope, the bioregional movement will spread, ultimately reencompassing the entire earth. Through
the bioregional reinvention of political geography human beings can
reestablish their lost affinity with nature and begin the desired return to
the primordial state of harmony.

Spatial Hierarchies

The eco-radical vision of a decentralized social existence meshes with a
longing for a nonhierarchical social order. Hierarchy, it must be noted,
has several distinct meanings in ordinary discourse. Primarily, it refers to
the social strata, differentiated by wealth and power, that exist in every
centralized state. Such social inequality is, of course, opposed by all
schools of radical environmentalism. A second connotation of hierarchy,
however, is of primary concern here; namely, the organization of geographical space into unequal, "nesting" economic and political sets.

Such geographical hierarchies are ubiquitous in all large-scale societies. In much of the United States, for example, several townships are
encompassed within a county, which in turn occupies a similar position
vis-à-vis the state, just as the state is similarly subordinate to the federal
government. Large corporations are similarly organized through regional
offices, smaller district offices, and so on. Similar spatial hierarchies
characterize legal systems, church structures, nonprofit corporations—
in short all organizations requiring extensive territorial organization
(Sack 1986).

To most eco-radicals all such forms of spatial hierarchy are anathema.
Indeed, in virtually all versions of contemporary leftist thought, hierarchy itself has become a term of opprobrium. Environmental radicals,
therefore must carefully insist that their envisaged bioregional con-
federations will be "natural" associations, based solely on horizontal affinities, rather than structured organizations relying on vertical
bonds. While advocating global thinking, most environmental extremists strictly deny the need for any social, economic, or political organizations existing at any scale larger than that at which direct person-to-
person contact can be maintained.

Urbanism

What then becomes of cities in the eco-radical future? Every city, and
indeed, every town, owes its existence both to the hierarchical organization of space and to trade. No urban area can provision itself, as all
necessarily draw sustenance from outlying rural zones. The relationship

between town and hinterland can be complementary, but it is never horizontal. The concentrated economic power of the city simply cannot be matched in the countryside.

Different urban arrangements found in different cultural milieus exhibit varying degrees of hierarchical organization. In modern industrial societies, however, vertical interurban ties are pronounced. Large cities exert financial and other forms of economic control over smaller cities, which in turn dominate regional towns and so on down the urban hierarchy (see, for instance, Wheeler and Mitchelson 1989). As Braudel (1981:481) writes, "a town never exists unaccompanied by other towns: some dominant, others subordinate or even enslaved, all are tied to each other forming a hierarchy, in Europe, in China, or anywhere else." Geographers debate the reasons why urban systems develop as they do (central-place theorists stressing retail trade [B. Berry 1967], mercantile theorists emphasizing wholesale trade [Vance 1970], and others positing more diverse origins [Conzen 1987]), but all would agree that equality among cities, let alone between cities and their hinterlands, is impossible.

A theoretically coherent, anarchic environmentalism would therefore reject urbanism outright. Many green extremists do just that, reflecting a deep prejudice against cities that has pervaded ecologistic thought since its beginnings (Bramwell 1989). Still, a certain schizophrenic attitude toward cities is evident in most eco-radical circles. While urbanism may be considered philosophically objectionable, many of the movement's adherents live in cities and admit to enjoying urban life. Some have even suggested that small to medium-sized cities might be incorporated within the bioregional fabric (Sale 1985:114).

Almost all eco-radicals, however, remain firm in regarding *large* cities as intrinsically destructive of both nature and humanity. The philosophy of primitivism certainly leaves no room for anything remotely resembling an urban concentration. But the mainstream eco-radical view is almost as uncompromising; Sale (1985:65), for example, regards the "contemporary high-rise city" as "an ecological parasite [that] extracts its lifeblood from elsewhere and an ecological pathogen [that] sends back its wastes." The eco-radical litigator Jeremy Rifkin again offers the most sweeping broadside against modern urbanism. To Rifkin, large, densely populated cities are little more than "high entropy" environments doomed to expire: "even as the city attempts to preserve itself," he informs us, "it actually fosters its own economic decline" (1989:173).

Humanistic eco-anarchists, who often favor small cities, remain equally hostile to the modern metropolis. Although Bookchin argues that "[c]ities comprised a decisive step forward in social life" (1989:180), a few pages earlier he complains that large cities generate a dehumanizing anonymity (1989:167). The latter sentiment is widespread in the eco-radical movement. Catton (1980:206), for example, believes that the density of modern cities creates a psychic overload that induces people to commit antisocial acts.

Because of this hostility to large, densely inhabited areas, urban-dwelling eco-radicals greatly fear increasing concentrations of population and economic activity in their own cities. Radical green activists of all stripes thus advocate limiting or even forbidding future urban growth (for example, Tokar 1987:66). Antigrowth proposals are often supported with a view to preserving existing city neighborhoods, pictured as small-scale islands of social sanity in death-dealing seas of urban anonymity.

Many deep ecologists further suspect cities of sapping the human spirit simply because they deny people their necessary contact with nature. When this intrinsic human need remains unfilled, as it must be in urban areas, social pathologies inevitably erupt. According to this view, human spirits can only be healed by a return to a natural landscape. Once this happens, many believe, social and ecological harmony may begin to be reestablished.

Ruralism Revisited

Suspicious of the city, dismissing industrialism, and disparaging trade, most radical environmentalists hope to recreate a rural, agrarian, society. Yet the dominant forms of rural life found in the industrialized world are roundly criticized as well. Contemporary American agriculture is singled out, with good reason, for being both ecologically destructive and dehumanizing (W. Berry 1977). Eco-radicals would much prefer a return to old-fashioned, animal-powered forms of cultivation that shun all chemical inputs.

Many green writers recognize that returning to small-scale, nonmechanized agriculture would require the labor of many more farmers than currently reside in the United States. As farmer-philosopher Wendell Berry (1981) argues, good stewardship requires a large number of cultivators to look after the land. Those persons joining the hoped-for exodus from the doomed and stultifying megalopolis could thus potentially find abundant employment as old-fashioned dirt farmers, a process that would simultaneously revitalize American agriculture.

■ The Liberal Rejoinder

The radical environmentalists' extraordinary faith in decentralized political power runs counter to the philosophies of both traditional liberalism and socialism. In the United States, movements espousing the devolution of political power, such as the various states rights campaigns,
have often been strategic ploys by the radical right to counteract reforming tendencies at the national level. On environmental as much as on
social issues, America's federal government has historically been more
forward looking than most local political entities. As Koppes (1988:240)
writes in regard to the history of the American conservation movement:
"conservationists often found decentralization frustrating for it tended
to reflect the immediate economic interests of powerful regional elites
rather than national priorities. Arguing that natural resources belonged
to the whole country, conservationists thus usually tried to have environmental policy made at the national level." Indeed, the main environmental agenda of the Reagan administration was precisely to shift
responsibility for environmental problems from the federal to the state
and local levels (Henning and Mangun 1989:75). While the rhetoric
associated with this move may have stressed the desirability of local
autonomy and freedom from meddling Washington bureaucrats, its over-
riding goal was nothing less than the gutting of environmental regulation. This is not to imply, however, that decentralization is always anti-
environmental. In certain circumstances a selective shift of authority
from the higher to the lower levels of a spatial hierarchy can in fact be
highly beneficial.

In recent years political-environmental theorists have carefully examined the ecological consequences of decentralization from the federal to
the state level. Several scholars advocating a federalist approach have
indeed discovered that certain American states often act as environmental pacesetters (Lowry 1992). Indeed, the national government has at
times attempted to weaken state-level pollution standards. But the federalist approach, stressing a carefully constructed balance of federal and
local (especially state) authority, must not be confused with the radical
decentralization advocated by green extremists. It is necessary to recognize, as Lowry (1992) demonstrates, that the ability of progressive states
to enact strong environmental measures is severely hampered whenever
interstate competition intrudes. In other words, in the absence of centralized coordination, pollution-generating firms can often thwart state
policy by departing, or threatening to depart, for less environmentally
sensitive jurisdictions. Even in economic sectors in which offenders

cannot relocate, such as agriculture, the lack of centralized authority will severely limit the diffusion of innovative control programs from the more progressive to the less progressive states. And finally, it must be recognized that some states will simply opt to "abdicate [environmental] responsibility altogether" (Davis and Lester 1987:563).

In industry and government alike, hypercentralization is indeed debilitating. Well-coordinated decentralization programs will thus be necessary in many spheres of power if a sustainable economy is to be devised. But it is equally imperative to recognize the many dangers inherent in the total dismantling of all centralizing structures. What is necessary, however, is a balance of powers among the local, the intermediate, and the central levels of authority. Coordinated decentralization, coupled with flexible centralization, must be the aims of a Promethean environmentalism.

This discrepancy between radical environmental theory (which automatically lauds the local and condemns the central) and the empirical evidence (which suggests that strong environmental legislation often originates in large-scale political entities, that even progressive instances of decentralization require some form of centralized coordination and that a great many local communities consistently oppose strict environmental measures) is completely overlooked by most radical greens. They would probably counter, however, that once we transform our ways of thinking and perceiving, obstacles to effective coordination as well as reactionary tendencies at the local level will vanish. The essential question is whether such a view can be justified.

Small Can Be Ugly
Small-scale communities are seldom as humane and ecologically sound as eco-radicals portray them to be. Opposing the green position here is an old tradition in social thought stressing the parochial, small-minded nature of both rural and small-town life. While the American intellectual tradition has often disparaged urbanism (K. Jackson 1985:68), to most intellectuals throughout history the city has been the scene of individual liberation rather than anonymous repression (Vance 1977:12). Local politics have often been pictured as dominated by grasping oligarchies, not by the equality-minded citizens' councils of eco-radical rhetoric. Significantly, antirural intellectual sentiments were stronger in an era when small-town and village life was fresh in most writers' memories; rural romanticization, in contrast, has flowered in recent years among writers who have grown up in comfortable suburbs, completely lacking direct experience of the small-scale social milieu. Longing for a

more meaningful social experience, American Arcadians long ago appropriated the venerable image of the New England village community as the democratic ideal. Unfortunately, even this utopian vision was in fact never much more than a "romantic literary conceit" (J. Wood 1991:46).

Some of the more thoughtful environmental philosophers acknowledge that small communities can stifle nonconforming or ambitious individuals (Daly and Cobb 1989:170). The dominant view, however, is that any such unpleasantries would vanish under a system of real participatory democracy. When all persons have equal and direct political power, many eco-extremists believe, a domineering elite could never arise. Similarly, the full diffusion of power would protect those individuals choosing idiosyncratic lifestyles. How existing local elites are to be divested of their power, and how existing social conventions and prejudices might be eliminated, are less often addressed.

In addition to the practical difficulties of instituting local egalitarianism, one must question whether small-scale participatory democracy would really eliminate social repression. Just as likely, the result would be a tyranny of long-winded individuals immune to boredom. Indeed, Anna Bramwell (1989:226) goes so far as to argue that participatory democracy is so wastefully time-consuming that members of eco-radical communes ironically "seem to need more of the earth's resources than other people." Or, as Oscar Wilde once reportedly quipped, "the problem with socialism is that it takes too many evenings."

To support the cause of direct democracy, eco-radicals have sought out historical instances of its successful institution. Unable to hold up their own or their forebears' experimental efforts in communal living, whose histories have typically been short-lived, they have turned instead to indigenous American social organization. One popular model of participatory democracy is the Iroquois Confederacy of seventeenth- and eighteenth-century North America (for example, Tokar 1987:13; Nollman 1990:13).

In fact, the Iroquois Confederacy is a particularly ill-considered exemplar. Admiring the Iroquois political system *of that era* for its democracy is akin to praising nazi Germany for its enlightened forestry. The Five Nations not only engaged in a highly successful campaign of ethnocide against their competitors in the fur trade, the Hurons, but they also raised the torture of war captives (those whom they chose not to adopt, at any rate) to an art. Victims were taunted while being slowly burned alive and having their flesh gouged from their bodies. Even small children were sometimes subjected to this treatment (see Sanday 1986:148; the recent Canadian film *Black Robe* provides a stunning visual portrayal of such practices). Direct democracy gives absolutely no guarantee of ethical social norms.[3]

A more far-reaching challenge, however, may be found in the human (or at least, human male) propensity for violence. One does not have to accept the Hobbesian proposition, that without a strong central authority a war of all against all is inevitable, to see that in the absence of some sort of peacemaking and law-enforcing mechanism, war, feuding, headhunting, and other culturally sanctioned forms of bloodshed have been dreadfully common. Large-scale political organization admittedly can lead to large-scale—and thereby more devastating—conflicts. The immediate point, however, is not to defend the existing geopolitical system (which is in need of reform), but rather to criticize the eco-radical alternative.

Cities and Environmental Enlightenment
While the dream of an anarchic rural utopia may be simply naive, opposition to urbanism per se is directly threatening to nature. As Paehlke (1989) carefully shows, urban living is in a great many respects far less stressful on nature than is rural existence. Given our current political economic structure (which, despite eco-radical hopes, is in no immediate danger of collapse), any movement of the American population away from cities toward the countryside will result only in a hastening of environmental destruction.

Urbanism's environmental benefits are most easily visible in the realm of transportation. Public transport, which is almost always less polluting than travel by private automobile, is feasible only in and between cities. The denser a city's population becomes, the more efficiently its public transport system can operate. Moreover, in urban core areas, walking is often the most convenient mode of travel. In America's countryside, in contrast, the automobile is generally the sole feasible means of transport. At present, rural Americans seem willing to drive ever greater distances to seek modern conveniences; small towns everywhere are decaying as their erstwhile shoppers cruise to the regional centers large enough to support shopping malls or, at least, discount stores.

The intrinsic energy efficiency of cities is evident in other aspects of life as well. Detached dwellings require far more energy to heat than do rowhouses, let alone apartments. Cogeneration, a process by which industries use what would otherwise be waste heat, is most feasible in areas of high density. More significant is the reduced energy costs of trucking goods from business to business and from business to consumer in the urban environment. Simply by virtue of its energy efficiency, the city pollutes far less on a per capita basis than does the countryside, given the same living standards. Noxious by-products may be more quickly diluted in rural environs, but the total output per person is generally much greater.

Equally revealing is a comparison of actual land use in the two environments. Urban dwellers typically require a small fraction of the space required by country people. This is readily apparent both in housing (multistory apartments versus detached, single-family dwellings) and in infrastructure (the more scattered the population, the greater the expanse of roadway). If America's present urban population were to be dispersed over the countryside, vast tracts of land would have to be converted to housing and transporting them, tremendously reducing wildlife habitat.

Radical greens may argue that if only city dwellers would move to underpopulated but overcultivated agricultural zones, wildlife habitat would not significantly diminish. This may be valid in theory, but one sees few back-to-the-land environmentalists relocating in northern Iowa. When Americans find themselves exhausted by urban living and inspired by the panegyrics of eco-radicals, they generally seek solace in forests, mountains, and sea coasts rather than in regions noted for their hog farms. As urban refugees stream into scenic landscapes, roads and subdivisions follow in pace.

One way to appreciate the environmental benefits of urbanism is to compare Japan and California, two areas of roughly the same land area with equivalent levels of economic development. Japan must support over 120 million persons, while California is home to only some 30 million. The Japanese population is highly concentrated in dense agglomerations; as many as 30 million persons reside in metropolitan Tokyo alone. California, on the other hand, is characterized by vast suburban sprawls of relatively low population density. Los Angeles County's 9 million persons, for example, are spread out over 4,068 square miles of land, an area larger than greater Tokyo. Moreover, most of California's rural zones, whether farmland, desert, or forest, have been experiencing rapid low-density growth in recent years. Due to outward migration from urban areas, a vast, vague, quasi-metropolitan corridor now spreads from San Francisco east as far as the Nevada border at Lake Tahoe.

The Japanese, not surprisingly, require far less energy on a per capita basis than do the Californians. Certainly population concentration is not the only reason for Japanese energy efficiency; a Spartan attitude in regard to residential heating and cooling, as well as vigorous investments in modern, energy-saving technologies, are also significant factors. In building a low-cost transportation network, however, population density is paramount. Sprawling greater Los Angeles, let alone California's vast exurbia of metropolitan refugees, could never support the efficient interlocking system of subways and commuter rail-lines that characterizes the Tokyo metropolitan area.

Even in such an unlikely area as wildlife preservation the Japanese are ahead of the Californians on several fronts. Japan, for example, supports a healthy population of *Ursa arctos*, the grizzly or brown bear. In California the same creature—the state symbol—was exterminated decades ago. Environmentalists (Promethean and Arcadian alike) advocate reintroducing grizzlies to California, but they are always shouted down with cries that the state is too crowded to support such a "wilderness species." Japanese brown bears have been able to survive in part because the country lacks a rangeland cattle industry (ranchers are usually the main enemies of large carnivores), but equally important is the concentration of its population in cities. Only the intensity of its urbanism allows Japan the space needed to maintain adequate brown bear habitat. If California had to support four times its present population, given its residents' aversion to urban settlement, it is doubtful whether any prime wildlife habitat would remain.

The Return to Nature and Its Destruction
The environmental destruction that would accompany ruralization on a massive scale is evident in the behavior of many back-to-the-landers themselves. Most well-intentioned, suburban-raised nature lovers quickly discover that nature itself conspires against their utopian plans for harmonious living. Insects, birds, and mammals soon begin raiding their gardens, while predators begin devouring their lambs and kids. Many neo-homesteaders do take pains to avoid harming their competitors; some California greens even purchase cougar manure, the scent of which will supposedly prevent deer incursions, to spread about in their vegetable patches. But many others, including some I have known personally, simply return to the behavior that humans have exhibited for millennia when faced with animals that threaten their livelihoods: they kill them.

In a great variety of circumstances eco-radicals seeking communion with nature are forced to compromise their ethics and mete out their own small forms of destruction. A precious example is provided by Jim

Nollman (1990) in a brief book called *Spiritual Ecology*. Nollman details the heroic efforts he undertook to keep deer out of his garden and insects away from his house. He had enjoyed modest success until an ant had the temerity to bite his young daughter. At this point, any sense of ecological responsibility seems to have evaporated immediately; Nollman promptly poured gasoline over the nest of the offending creature and set it ablaze.

The philosophy of deep ecology, especially the primitivist variant, is threatening to wildlife in a more direct manner. Since predation is an inescapable aspect of nature, these self-styled protectors of the wild believe that human beings should feel no qualms about living a predatory existence. In fact, a glorification of hunting runs deeply through their own literature. One of the movement's heroes, Aldo Leopold (1949 [1966]:129), found no greater joy than in slaughtering waterfowl on frosty autumn mornings: "I cannot remember the shot; I remember only my unspeakable delight when my first duck hit the snowy ice with a thud and lay there, belly up, red legs kicking."

As a nonradical environmentalist, I do not necessarily oppose hunting. Where predators are absent, it is often necessary to cull fecund herbivores lest they degrade their own habitats. More importantly, a coalition between hunters and environmentalists is absolutely vital if habitat is to be preserved. Yet I can only shudder on hearing an avowed environmentalist boast of "unspeakable delight" in the petty act of slaughtering another living being.

Unless either human numbers are extremely small or some form of wildlife management (whether technical or traditional) is instituted, hunting *is* inherently destructive. The close-to-nature American pioneers eliminated virtually every large mammal remaining in most parts of this country. Only when scientific game management was developed near the turn of the twentieth century did certain game species, such as the whitetail deer, begin to rebound. Yet many deep ecologists denounce the concept of professional wildlife management as an arrogant affront against the natural world.[4] Nature will manage itself, they believe, and if only we would become a part of the natural world we would never have to fear the consequences of our own acts. Many seem to believe that if hunters "ask forgiveness" of the animals they kill, all will be well. This did not work for the Rock Cree, nor will it work for our own neo-tribals. Wildlife management as presently practiced does deserve criticism for its focus on game species, but only careful wildlife oversight can make hunting a tolerable activity. Otherwise, the more Earth First! warrior-hunters we have prowling our forests, the more wildlife will suffer.

The Political Geography of Ruralization
The ruralization of the American populace would also appear to threaten nature insofar as it might undermine electoral support for environmental reform. American urbanites consistently support environmental measures to a far greater extent than do rural dwellers. As Henning and Mangun (1989:8) argue, rural values are simply more utilitarian than urban values. Neither does this seem to be merely a measure of the residual conservatism of traditional small-town life. Even urban refugees show a marked propensity to ignore environmental problems as soon as they leave the polluted, traffic-snarled city. Calaveras County, California, a rapidly growing green refuge on the west slope of the central Sierra Nevada Mountains, provides a good example of such tendencies.

The population of Calaveras mushroomed after 1970 as immigrants fleeing the congestion of the Bay Area and greater Los Angeles began to arrive, growing from 13,700 in 1970 to 31,000 in 1988. Newcomers now outnumber natives by a substantial margin. But the county continues to vote solidly against every conservation proposition on the California ballot. The spring 1990 Passenger Rail and Clean Air Bond Act (Proposition no. 108), for example, received 42 percent of the vote in Calaveras; San Francisco, in contrast, awarded it 75 percent of its vote, while even hyperconservative Orange County gave it 51 percent of its vote. Overall, the correlation in California's 1990 elections between population density and support for environmental protection is profound, as can be seen for three representative issues in tables 1 and 2.[5]

Other political-geographical studies have uncovered similar patterns. Greenburg and Amer (1989), for example, show that in New Jersey, urban dwellers, both rich and poor, were far more likely to vote in favor of a bond issue to clean hazardous waste sites than were rural residents. Overall, in the United States at least, the higher the concentration of humanity, the greater the electoral concern for nature.

In casual conversation the reason many residents of Calaveras County give for voting against the environment is simple: "we don't have traffic and pollution problems up here, so why should we pay to solve them— let the city people clean up their own mess." This attitude seems to be nearly as common among those who recently moved to the mountains, precisely to enjoy its pristine landscape, as it is among long-term residents. As table 2 shows, rural counties in California that have experienced population explosions in recent years—fueled almost entirely by outmigration from urban areas—vote consistently against environmental propositions of all kinds. Here one sees the variety of local environmentalism that is actually produced by the eco-radical agenda of total

Table 1   Population Density and Support for Proposition 128 (Environment Public Health Bonds, or "Big Green," November 1990; Northern California Counties Only) (Vote in Percentages)

Population Density (Aggregated by county)	Yes Vote	No Vote
Counties containing:		
Over 2,000 persons per square mile	62	38
1,000–2,000 persons per square mile	39	61
200–1,000 persons per square mile	36	64
50–200 persons per square mile	29	71
10–50 persons per square mile	20	80
Less than 10 persons per square mile	22	78

decentralization and local autonomy. Only when outside agents seek to dispose of toxic wastes in places like Calaveras do their residents take on environmental colorings. In the face of such threats, not surprisingly, their pro-environment response is rapid and overwhelming.

Urbanism Reconsidered

Many Americans (myself included) cannot abide life in an urban environment and will continue to reside, at least on weekends, in rural or quasi-rural areas. There is nothing intrinsically wrong with such a personal decision, and a Promethean environmentalism would never advocate forced urbanization. But it is equally imperative to realize that another sizable segment of the populace cannot tolerate rural existence and in fact craves city life. Any attempt to demolish the framework of contemporary urbanism would be simultaneously an attack on human freedom and dignity.

Yet few eco-radicals, it would appear, would be willing to tolerate the life-style proclivities of urbanites. Regarding major cities as wretched pits of technological devastation, extremists like Rifkin argue that massive agglomerations of humanity must simply vanish if the planet is to survive. Few realize the totalitarian implications of such thinking, perhaps because it does not occur to them that anyone would actually choose to live in such a degraded and degrading habitat. The Manhattan and San Francisco housing markets provide ample testimony, however, that a good many Americans want to live in *densely* concentrated urban environments. Indeed, the social sterility of many American cities re-

Table 2   Environmental Voting Records of California Counties with Population Densities under 50 Persons per Square Mile (1989) Whose Populations More than Doubled between 1970 and 1988

A. Proposition 108 (Passenger Rail and Clean Air Bond), June 5, 1990 (Vote in Percentages)			B. Proposition 117 (Wildlife Protection), June 5, 1990 (Vote in Percentages)		
County	Yes Vote	No Vote	County	Yes Vote	No Vote
Alpine	48	52	Alpine	49	51
Calaveras	42	58	Calaveras	41	59
Lake	38	62	Lake	39	61
Mariposa	41	59	Mariposa	42	58
Mono	47	53	Mono	52	48
Tuolumne	46	54	Tuolumne	37	63
Statewide	56	43	Statewide	52	48
Counties with more than 1,000 persons per square mile (Northern California only)	61	39	Counties with more than 1,000 persons per square mile (Northern California only)	55	45

sults partly from the fact that they are not crowded enough to support a truly urban, pedestrian way of life, offering instead only a vast collection of suburbs centered around a small collection of lifeless high rises and strung together by congested freeways.

In other industrialized societies the attractiveness of large urban centers is even more pronounced. In the United States most cities do suffer horrendous social ills that force many to flee to the suburbs, but this reflects America's unique social pathologies, not the urban condition per se. Japanese, European, and Canadian cities are not particularly violent places. Nor is city life necessarily dehumanizing. Anyone who thinks that urban pressures make people rude and short of temper is advised to visit Tokyo. Nor is it entirely coincidental that the livable cities of Japan and Europe are characterized by much higher population densities than are the depopulating American cities so despised by eco-radicals.

An environmentally sound society should *encourage* the growth of high-density urban centers, cities in which residential, commercial, and industrial functions are closely configured. But any concerted movement

toward urban intensification would be strongly countered by neighborhood activists, individuals who mistakenly believe that in opposing
development they are protecting the environment. Urban preservation
movements can be a positive force, especially when they fight to maintain residential districts that would otherwise be converted to less intensive land uses. But at its worst, neighborhood environmentalism becomes a parochial movement more concerned with the availability of
parking spaces for the upper middle class than with the ecology of the
planet. Acting locally in this case is coupled not with thinking globally
but rather with considering above all one's own convenience.

The demographic trajectory of the United States presents anti-urban
eco-radicalism with a serious dilemma. The population of the United
States is presently growing at roughly 1 percent a year, an increase due
almost entirely to immigration. As would-be progressives, eco-radicals
can hardly call for migration restrictions without appearing racist, yet as
nature lovers they abhor the kind of development necessary to provide
newcomers with jobs and housing. With the notable exception of the few
primitivists, who firmly advocate closing America's borders, few have
faced this contradiction. Regardless of anyone's stance on immigration,
the population of the United States will continue to grow for the foreseeable future. Urban areas have the potential to absorb most of the increase, but only if abundant, high-density housing is provided. Where
this is politically unfeasible, continued suburbanization will be the only
alternative.

Similarly, as population mounts, new commercial developments will
be essential if we are to avoid entering a period of per capita economic
decline. Since urban environmental activists typically oppose commercial developments even more strongly than residential construction, a
strong impetus is provided for firms to locate new office complexes far
into the suburbs. This trend, already underlain by land price differences
and especially by the desire to find a compliant "pink-collar" workforce
(Nelson 1986), will result in the continued geographic expansion of low-
density metropolitan zones. As back offices move out of urban cores, the
suburban fringes around many cities are swelling to encompass extensive, formerly rural tracts. In northern California the rich agricultural
districts of the northern San Joaquin Valley are now being paved over to
form bedroom communities for San Francisco's office-park suburbs of
Pleasanton, San Ramon, Walnut Creek, and Concord. The environmental insanity of mass suburbanization is evident to eco-radicals. What
they fail to realize is that their own philosophies and actions do nothing
but encourage it.

The Environmental Catastrophe of Suburbanization
Environmentally, suburbs may be the worst of all possible worlds.[6] Residential density here is low enough to preclude effective mass transport, yet high enough to exclude all but the most adaptive species of wildlife. Suburban planners and dwellers usually further degrade their environments by building unnecessarily wide streets (often mandated by well-meaning city codes in the name of fire protection), by paving broad areas for driveways, sidewalks, and patios, and by devoting enormous sums of energy, water, and toxic chemicals to maintain a vegetative cover of exceptionally low biological diversity. Most well-to-do suburbs actually embrace a form of landscape totalitarianism in which a person can be fined for erecting a clothesline or for allowing a diverse assemblage of wild plants to grow.

Yet the suburbs spring from exactly the same prejudice against cities that has pervaded the writings of eco-radicals for the past century. The original notion was that in a bucolic environment, where each family could enjoy its own home and garden, the close presence of nature would cure the stress-induced diseases of urban life and in so doing restore humanity to a more blissful state (K. Jackson 1985). If the suburbs have failed in this civilizing mission, the more radical ruralization advocated by today's green extremists would likely have even more devastating consequences.

As environmental historian William Cronon persuasively shows, the rural-urban dichotomy that has infused so many American cultural myths and that has sustained eco-radicalism ultimately hides more than it reveals. Whether praising the city or disparaging it, partisans of both the urban and the rural have incorrectly assumed that the two "are separate and opposing worlds, that their divisions far outweigh their connections" (Cronon 1991:17). But this habit of regarding the city and the country as distinct and separable has "obscured their indispensable connections. Each [has] created the other, so their mutual transformations in fact [express] a single system and a single history" (Cronon 1991:368).

The Fantasies of Decentralization
The eco-radical fear of cities and other large-scale structures seems to be rooted in an underlying belief that there are really only two basic forms of social organization: centralization, which implies totalitarianism, and dispersion of power, leading to social harmony and, ultimately, utopia. Intermediate forms are implicitly considered unstable, tending eventually to yield to centralized control.

This species of dualistic thought is unhelpful for devising a sustainable socioeconomic order. A degree of centralization is essential if we are to avoid economic chaos, internecine strife, and localized political repression. To accept limited central authority, and the attendant necessity for hierarchical organization, does *not* imply condoning the brutal social hierarchies associated with precapitalist states, certain capitalist economic organizations, or contemporary totalitarian countries. To the contrary, a happy medium between anarchy and centralization is not only possible but highly desirable.

Human organizations appear to run most smoothly when they are characterized by flexibly constituted socio-geographical hierarchies. A local community should enjoy a limited autonomy that is, in crucial areas, superseded by a higher level of authority. Such flexible centralization (or coordinated decentralization), sometimes called federalism, is premised on the notion that local communities, larger regional groupings, still larger "states" or provinces, and finally a "federal" government, should all have their own spheres of power. If a correct balance is struck, there will be no inherent tendency for increasing centralization, although certain political interest groups may well pull for it.

Hierarchy and Globalism
Hierarchy is, in the final analysis, an inescapable principle of organization itself. Arthur Koestler (1978:31) is not alone in noting that, "all complex structures and processes of a relatively stable character display hierarchic organization, regardless of whether we consider galactic systems, living organisms and their activities, or social organizations." Even most tribal societies are generally organized in such a fashion. As a noted Africanist observes: "It has often been remarked by ethnographers . . . that African cultures are suffused with a sense of hierarchy in social, political and ritual relations. . . . This holds true even for those 'segmentary' or 'acephalous' or 'stateless' African societies that are sometimes labelled (or rather mislabelled) as 'egalitarian'" (Kopytoff 1987:35). Social hierarchies of some sort are intrinsic to virtually all forms of social organization, whether capitalist or communist, Western or Eastern, industrial or agricultural.

Political and economic hierarchies, to be sure, do entail inequality. A supreme court justice presides over a circuit court judge, and a corporate division manager must have greater authority than a local manager. But variations in the scope of individual power do not necessarily imply relations of domination, as many radical environmentalists seem to believe. Domination is a symptom of system pathology, not a standard mode of hierarchic structure.

This antipathy to hierarchy is not unique to eco-radicalism but rather endemic to the entire far left, and as such it is deeply ingrained within American academia. Although the antihierarchic turn, motivated by an admirable commitment to equality, is probably just naively idealistic, some writers argue that it is actually threatening to democracy. As David Lehman (1991:100) explains: "The demolition of hierarchies is pursued as an end in itself, leaving us nothing to hold onto, nothing with which to resist the imposition of a new hierarchical order. The twentieth century's dismal chronicle of new hierarchies, established upon revolutionary new theories, does not make one welcome the prospect."

The world's greatest political failing, I would argue, lies not in the existence of spatial hierarchies, but rather in their lack of depth. Our prevailing hierarchical structures are severely truncated at the top, and, in many parts of the world, at the bottom as well. The main provision for global power, the United Nations, is an anemic body based on the questionable assumption of equivalence among all sovereign polities, where a country of 100,000 inhabitants is awarded the same weight as one of 1 billion (with the notable exception of those few selected for the security council). Of course, the UN does offer an important forum in which the first steps toward coordinated global environmental protection may be devised (MacNeill et al. 1991). But in the end, a more efficacious form of planetary power may be necessary if we are to institute appropriate remedies. Fortunately, the governments of several European countries are now beginning to think along these lines (Porter and Brown 1991:153).

Membership in a strengthened UN or some other variety of global federation need not entail the surrender of independence. Only when states are violating essential human rights or natural protection laws (for example, allowing trade in endangered species) should global authority be invoked. But even conventional pollutants often transcend national boundaries, while others, such as carbon dioxide and chlorofluorocarbons (CFCs), threaten the entire planet. While patchwork agreements can generate band-aid responses to the most serious threats, they are ineffective in devising long-term solutions. Systematic global cooperation, requiring a political body much more powerful than the current United Nations, may well be necessary if we are to begin addressing the most serious planetary dilemmas.

A first step for a newly constituted global body might be the suppression of the multitude of small-scale but vicious wars that currently plague several of the most impoverished regions of the world. As long as Sudan is engaged in a brutal civil war, both its people and its environment will continue to suffer horribly. The world over, war marches hand in hand with ecological destruction and economic privation. Just as the emergent states of late medieval and early modern Europe outlawed private warfare and in so doing vastly enhanced their own potential for economic growth, so too a global confederation could achieve the same result at the planetary level.

One of the major challenges to instituting world peace is the fact that most contemporary wars are fought not between independent countries but rather between competing ethnic groups located within the same national boundaries. Here overcentralization is indeed the most common problem. Regardless of what we may call them, few Third World countries are nation-states. They are rather melanges of culturally distinct peoples forced into single polities by their former colonial rulers (a nation, in contrast, is by definition a people self-consciously united). While Western scholars and governments have long urged newly independent countries to "build their nations," this has usually entailed the mass assault by central governments on nonconforming local traditions and institutions. In this struggle, the United Nations has actually proved to be a bulwark of state repression against nonstate, nonconforming nations (van den Berghe 1981:3; Knight 1983:128). Threatened with cultural extinction, minority groups not uncommonly take up arms to try to maintain a measure of self-determination. Encompassing states generally respond with quick brutality (Nietschmann 1987).

Examples of attempted ethnic centralization in postcolonial regimes, followed by mass repression and desperate rebellion, are legion. The process is perhaps most clearly evident, however, in the Buddhist-socialist state of Burma, which now styles itself as the Union of Myanmar. Over the past twenty years the Burmese army has engaged in a campaign of genocidal brutality against the Karen, the Chin, the Naga, the Shan, the Katchin, the Rohingya, and the Mon peoples (see the various articles in *Cultural Survival Quarterly* 13, no. 4, 1989). Partly as a result, the Burmese economy has stagnated for decades. Moreover, in order to maintain its grip over this crumbling state, the Burmese military has carried out extraordinary repression even against members of the dominant Burman nationality.

It is highly revealing that Burma was E. F. Schumacher's model for a benign system of Buddhist economics in *Small Is Beautiful*. While Schumacher's book was written before the Burmese debacle had fully come to light, such an excuse cannot be granted to eco-radical John Young, who evidently continues to regard Burma as a virtuous LDC (lesser developed country) that has "some useful lessons for the rest of us" (1990:115). The lessons of Burma, I would retort, may be powerful, but they are all negative.

A potential solution to such national struggles would be the creation of spatial hierarchies of greater flexibility. Local communities of distinct cultures must possess a substantial degree of local autonomy. But such units need not be politically independent; instead, they might be federated with other similar units. Such encompassing federations, in turn, could join with similarly constituted neighboring organizations. Ultimately, in many areas of the world, the independent state itself may cease to be the foundational entity of political affiliation.

While the demise of the nation-state may seem a hopelessly utopian dream (or a terrifying nightmare, depending on one's perspective), some evidence suggests that precisely such a process is beginning to occur in the very heartland of that geopolitical form. Contemporary Western Europe is experiencing pressure for political devolution (the Scots may yet opt out of Britain; the Basques and Catalans may well quit Spain) at the same time that much of the subcontinent is moving toward flexible unity through the European Community (EC). Moreover, economic networks are arising in Western Europe that unite regions located in different countries; most notable here is the "four motors agreement" linking Lombardy, Catalonia, Rhône-Alpes, and Baden-Württemberg (Murphy 1991:11). Even the staid journal *The Economist* has recently wondered whether we may soon be saying "goodbye to the nation-state" (June 23, 1990). As geographers John Agnew and Stuart Corbridge (1989: 273) see it, the contemporary nation-state is, in many instances, both too large for purposes of social identity and too small for effective economic integration. Whatever shape the planet's future political geography may take, we can expect that it will involve some striking departures from the contemporary jigsaw map of two-dimensional, theoretically equivalent, sovereign states.

Geographical Advantage and Autarky

Contrary to the eco-radical vision, economic integration through extensive trade networks is not only beneficial for economic development but is also essential for future ecological health. Without the specialization made possible by transregional economic connections, and without the ability to transport essential resources over long distances, our entire economic and technological edifice would collapse. While most radical environmentalists claim that they would welcome the resulting global economic chaos, its consequences would both impoverish the surviving communities and wreak ecological devastation. The local autarky envisaged by the greens would force us back to something very similar to the early medieval economy, one in which the vast majority of people lived in dire poverty. In fact, even in early medieval times, more long-distance trade was conducted than eco-radicals would care to allow. As Fernand Braudel (1990:121) writes in regard to Carolingian France: "It must be understood once and for all that no economy of any size could have survived under the mortal regime of autarky."

Nutrition standards would also decline sharply in a fully decentralized world. According to the eco-radical vision, every community, or at least every bioregion, would have to produce its own food. For instance, the inhabitants of coastal Oregon would need to cultivate wheat or rye in order to eat bread, a challenging endeavor along that fog-drenched coast. Although they could subsist on a more climatically appropriate staple such as the potato, a true deep ecologist might object that this tuber, as an Andean crop, is not native to the bioregion. In fact, no high-yielding food crops are native to this area, and it is doubtful whether salmon, the one-time staple of the region, could ever return in adequate numbers. Perhaps coastal Oregonians could subsist on bracken fern roots, as once did the Maori of South Island, New Zealand; I invite them to try.

One might also wonder how a future bio-regional coastal Oregon would harvest solar power, considering how little sunshine actually falls on its fog-shrouded shores. It certainly would not want to import solar-generated electricity from sunnier climes, since this would undercut the self-sufficiency that eco-radicals prize above all else. Obviously, with a cessation of transregional economic connections, ecologically benign technologies would be damaged just as severely as destructive ones.

The Promethean environmental viewpoint, in contrast, holds that it is both economically and ecologically sensible to generate solar electricity where the sun shines and to gather wind energy where breezes blow. The simple principle of geographical advantage, I would argue, should be exploited both to minimize energy and resource waste and to make human life more pleasant. Yet all notions of geographical advantage, comparative or absolute, are disparaged by radical environmentalists, individuals who hold an unreasonable prejudice against trade in all forms.

Global Enterprise and Local Cultures

The eco-radicals' fear that global enterprises will demolish local traditions, replacing them with a Disneyesque mockery of true culture, is not only misguided but deeply arrogant as well. Who is to be the global cultural arbiter, deciding which American films or records are unsuitable for foreign eyes and ears? Eco-radicals, many of whom would evidently endorse worldwide censorship on a massive scale, entertain severely misplaced fears. Indigenous societies the world over *are* gravely endangered, but such threats stem from nation-building central regimes, wealthy land-grabbers and forest clearers, and, in places like Bangladesh's Chittagong Hills, hungry, landless peasants from adjacent areas of high population density. Neither Madonna nor Sylvester Stallone has ever destroyed a culture.

In fact, throughout the world, local peoples are able to incorporate Western trappings without compromising their own cultural systems. The largest traditional, prestige feast I witnessed in highland Luzon, one that entailed feasting some 5,000 persons at the cost some $15,000, was celebrated by a family holding the local Coca Cola distributorship. Moreover, the majority of guests in attendance were reasonably conversant with the latest products of the global entertainment industry. Nor are such interests anything new. Since the early years of this century, even casual travelers have discovered that the Kankana-ey people enjoy country music, western films, and cowboy boots; in local parlance calling someone a cowboy is a great compliment. But there is no evidence that such superficial borrowings have had any deleterious effects on their core cultural values and social institutions.

I would further suggest that the excoriation of multinational corporations is misguided and potentially dangerous. Damning firms whose productive operations span national boundaries is not only parochial but may also feed reactionary forms of nationalism. To forbid economic endeavors to cross political divides, whether by discouraging trade or by disallowing the establishment of foreign subsidiaries, would only strengthen the division of humankind into contending national blocks. Such nationalism has proved itself many times over a potent force for both war and repression.

The current direction of corporate evolution appears, in any event, to be headed away from classical *multi*national organization toward truly *trans*national structures. A prime example is the electrical engineering giant Asea Brown Bovari, which is at once Swiss and Swedish, tempered by a strong North American component (the latter acquired with the purchase of Combustion Engineering). Of special significance for the present discussion, 15 percent of ABB's sales now come from pollution control equipment, and the company is poised to make the greatest corporate contribution to the cleanup of Eastern Europe (*Business Week*, July 23, 1990). Strategic link-ups between companies based on different continents also indicate a growing internationalization of capitalism. Ford and Mazda, for example, are now complexly intertwined at every level, from engineering and production to finance and marketing.

As liberal political economist Robert Reich (1991) argues, we may be entering an era in which major corporations will cease to have any national affiliation whatsoever. In Reich's controversial vision, corporations are seen as becoming flexible, cosmopolitan networks composed of semiautonomous deal-makers ("strategic brokers") whose various operations can leap about the globe in a restless search for greater efficiency. Not only do national ties lose their importance in such a world, but hierarchical centralization begins to diminish as well: "Instead of a pyramid, then, the high-value enterprise looks more like a spider's web" (Reich 1991:89). Ownership and control likewise grow ever more diffuse, as value is increasingly controlled not by investors but rather by the symbolic analysts who possess the skills and ideas necessary for creating and marketing products and services.

Reich's thesis is no doubt overstated, but the internationalization of capitalism is an unmistakable phenomenon. The transcendence of national boundaries by corporations is clearly a two-edged sword. On the positive side, we might expect international animosities to be quelled as economic integration proceeds. The problem, however, is that social control is not easily applied to a corporation with dispersed production centers, much less to one without a single administrative core. If too much pressure for environmental cleansing is placed in one region, a firm may simply relocate its operations to more congenial political environments. There are also social hazards; "the darker side of cosmopolitanism," Reich (1991:309) informs us, is that elite "world citizens" may never "develop the habits and attitudes of social responsibility." To my mind, such potential problems only strengthen the argument made above in favor of stronger global authority.[7]

Corporate Centralization

As Reich's analysis suggests, there is little evidence available to support the allegation that capitalist firms are growing ever more centralized. As many business theorists have noted, hyper-centralization is not conducive to efficient production. True, in the late nineteenth century and early twentieth century, increasing central power was a necessary strategy in new, technologically oriented industries. Firms in countries such as Great Britain that resisted the trend began to stagnate as a consequence (Chandler 1990). But the period of agglomeration in business seems to be ending. Indeed, the most savvy marxist analysts typically label the present era as one of "disorganized capitalism" or "flexible accumulation" (Scott and Storper 1986; Harvey 1989).

To regard large corporations today as massive, undifferentiated economic leviathans is highly inaccurate. Overly centralized corporations tend to ossify and become unresponsive. The more successful multinationals usually delegate a good deal of authority to their subsidiaries and branch operations. Another current trend is that of the subsidiary spin-off, as companies learn that it is often more profitable to concentrate on their areas of core competence. America's high-tech industries, in particular, continue to generate numerous new corporations, not all of which are destined to be swallowed up by larger firms. Other business trends also point away from increasing centralization. The average size of new plants opened in the United States, for example, has been shrinking for many years (Knox and Agnew 1989:369). Moreover, through the 1980s small firms accounted for most of this country's employment growth, while large companies actually shed employees in great numbers (Knox and Agnew 1989).

None of this means, however, that the United States has entered an era of rampant economic decentralization. Centripetal and centrifugal forces exist side by side in unstable tension. Capitalism perpetually seeks out the most efficient modes of operation, and while these vary greatly from place to place and era to era, they certainly will never entail either total centralization or complete decentralization.

The same processes are also at work in other countries. Americans often think of the Japanese economy as dominated by a handful of technologically adept giants. While this view is accurate in important respects, Japanese capitalism equally depends on a multitude of small subcontracting firms. Moreover, due to small-batch technology, worker shortages, and the much-vaunted "just-in-time" delivery system, small Japanese corporations are increasingly able to graduate from subcontracting and begin competing with their large-scale patrons (see *The Economist*, October 27, 1990). And even at the large end of the Japanese industrial scale, conglomerates such as Mitsubishi owe their vast power less to central control than to complex networks of horizontal affiliation. The *keiretsu* system of allied but competing firms united through interlocking directorates—exemplified by the Mitsubishi group—may well represent capitalism's next wave (*Business Week*, September 24, 1990; Eccleston 1989). There is even some evidence that a modified American *keiretsu* system is developing, exemplified at present by IBM and Ford, both of which are linked increasingly closely to their many suppliers (*Business Week*, "Learning from Japan," January 27, 1992).

A different permutation of decentralized production may be found in another thriving capitalist region, north central Italy's Emilio-Romagna. Here medium and small firms relying extensively on subcontracting have proved a formidable force in global competition (Porter 1990). Interestingly, many of the area's postwar entrepreneurs had been socialist artisans who set up their own shops after a series of prolonged strikes in the 1950s (Holmes 1986:88). Given the dynamism of such variant forms, the specter of a few gargantuan companies someday exercising a stranglehold on the world economy increasingly appears to be little more than an figment of eco-radical paranoia.

Many contemporary technological innovations also lend themselves to increased decentralization. This tendency is pronounced in the realm of electricity generation, a focal point of eco-radical concerns. As Ged Davis (1990:60) explains: "Technology is also mediating a shift away from large, centralized power plants to smaller decentralized ones. Improvements in electronic communications, control and computing technology have made it easier to monitor and regulate complex grids remotely. With the arrival of new gas turbines, small engines, solar cells and other technologies, the economies of scale . . . are diminishing." Similarly, in steel production, the "mini-mill" revolution of the past decade has also led to increased decentralization (Dertouzos et al. 1989: 286).

In the critical field of telecommunications, the tension between centralizing and decentralizing functions is more complex, giving rise to a heated debate among scholars (Moss 1988). But if we look at the conjunction of governmental policy and technological change, it becomes clear that the American industry as a whole has experienced substantial decentralization. Not long ago there was no option but to communicate over the lines of a single leviathan. The break-up of AT&T and the rise of new long-distance carriers brought increased competition, but only with the more recent emergence of alternative access networks have local options emerged. Using a combination of dedicated lines, leased channels, and microwave facilities, such alternative systems may well become formidable competitors to the baby Bells. The idea of a telecom monopoly is now "as outdated as the copper-wire technology that [once] prevailed" (Business Week, March 25, 1991, p. 96). Cellular telephones, as well as personal communications networks, offer another new communication route, and with the recent FCC approval of public service by SMR (specialized mobile radio) firms, mobile telecom is also growing less centralized and more competitive.

Contrary to eco-radical theory, the rise of telecom alternatives and the development of increasingly sophisticated technologies is making surveillance *more* difficult, not less. Indeed, cellular telephony in its early days acquired an unsavory reputation as the favored communication channel for drug dealers wary of tapped phone lines. While airwave transmissions can be monitored, improved signal scramblers make this ever more challenging. Legislation has also offered protection against surveillance, whether by private or governmental parties. The 1988 Electronic Communications Privacy Act, in particular, has helped safeguard the citizens' right to conduct unmonitored conversations. Certainly our rights to privacy are never secure, but such threats as do exist are located in the political, not the technological, sphere.

Firms Large and Small

Just as hyper-centralization has proved destructive, so too unmitigated decentralization is not without its hazards. While the development of monopolies must be continually guarded against, the parcelization of the economy into a multitude of tiny firms is potentially just as worrisome. Dertouzos and the other members of the MIT Commission on American Productivity (1989:55), to cite but one example, argue that the continuing spinning-off of high-tech firms may actually prove debilitating if carried to an extreme. Whatever the direction of corporate evolutionary trends, the time has come to reassess the eco-radical—and indeed, the traditional American liberal—preference for small firms.

Much empirical evidence suggests that the eco-radical picture of large companies as intrinsically more exploitative and dehumanizing than small ones is simply groundless. Brown, Hamilton, and Medoff (1990) demonstrate in unambiguous terms that small companies on aggregate offer their employees lower wages, fewer benefits, and weaker safety standards than do large firms operating in the same sector. More significant for the present discussion, small businesses much more easily avoid compliance with environmental regulations than do their larger competitors. The implications are clear. If large organizations are consistently penalized and small firms abetted, as radical greens urge, we could expect more workplace accidents and far greater emissions of industrial pollutants.

The Bioregional Fallacy

The eco-radical critique of scale in modern industrial society is deeply flawed, based as it is on a misunderstanding of hierarchic organization and on an outdated view of corporate structure. Its theoretical basis for small-scale social reconstruction is, on the other hand, hopeless. The bioregion is, to put it most bluntly, little more than a construct of bad, outdated geography, one in which the region itself is consistently mystified (Alexander 1990). Geographers labored for decades attempting to find nonarbitrary, uniform regions based on a correspondence of human and natural features, but eventually they abandoned the quest as pointless (Hartshorne 1939; Minshull 1967). Instead, most now concur that the various attributes of a given region all have their own stubbornly individual geographies, usually exhibiting little spatial correspondence (M. Lewis 1991). Ecologists have similarly discovered that one can never isolate a coherent ecosystem or even a plant community, since ecosystemic exchanges are typically continuous over geographical space and since each individual plant species exhibits a unique distributional pattern. What may appear as a firm linkage, as for example between the piñon pine and the juniper in a large portion of America's southwest, is merely a reflection of the fact that the two species' ranges have a considerable zone of overlap.

A geographer or ecologist can still delineate regions or ecosystems, but the careful scholar will remember that they are always to a large extent arbitrary. Some geographers have gone so far as to call all regions "artistic creations" (Whittlesey 1954), while Brennan (1988:121) reminds us that ecologists "split up the world" in a "rough and ready way." Such rough creations are not without utility; the International Union for the Conservation of Nature's (IUCN) division of the earth into 193 "biogeographical provinces," for example, has proven useful for devising conservation strategies. But each is certainly not a given in the sense implied by most bioregional advocates.

Admittedly, some bioregionalists have also recognized the difficulty of delineating nonarbitrary regions. Sale (1985:55), for instance, argues that the ecologically attuned practitioner must feel or sense the contours used in constructing natural boundaries. Since they prefer intuition to scientific precision, few eco-radicals would regard such difficulties as in any way threatening to the bioregional project. But if bioregionalism were ever to move from the realm of utopian dreaming to that of actual implementation, one could expect that serious dilemmas would emerge in the allocation of "sensed" bioregional space.

Even if we accept the utility of delineating more or less arbitrary natural regions, the question of whether human communities should, and indeed can, be organized around them raises a separate set of issues. Most eco-radicals argue strongly in the affirmative. Believing that a correspondence of biological and cultural regions was the norm among preindustrial societies, proponents contend that a bioregional mode of existence is both feasible and highly desirable.

Careful geographical analysis reveals the inadequacy of this formulation. Sale's (1985:60) own example of the "tribal conglomeration of Algonkian-speaking peoples" corresponding to the "ecoregion" of the "Northeast hardwoods" is meaningless. First, the name northeast hardwoods is not a recognized ecological region. Even if one were to accept a climactic vegetational model, one would still have to differentiate a beech-maple community from an oak-hickory community in this area. Moreover, both of these vegetational associations extended much farther westward than the traditional Northeast. Finally, many of the indigenous residents of this region were Iroquoian- rather than Algonkian-speaking, while Algonkian speakers were by no means confined to this environment (many lived in the subarctic, while others, such as the Cheyenne, eventually inhabited the Great Plains). A similar critique could be made of Young's asscrtion that medieval European kingdoms often corresponded to biologically defined regions. One has only to examine the shifting boundaries and vague definitions of such polities (for example, McEvedy 1961) to discover how nonsensical this notion is.

It is true that in pre-Columbian North America, and other parts of the world as well, broad cultural complexes were associated with broad natural environments. The various peoples of the Pacific Northwest, an area characterized by a mild, rainy climate, supporting dense coniferous forests, and blessed with abundant salmon runs, shared a wide variety of institutions and cultural traits (although such commonalities did not include such essential aspects of culture as language). But a sharing of lifeways did not provide these people any sense of bioregional solidarity; Northwestern communities frequently engaged each other in vigorous battle.

Kirkpatrick Sale, the most noted advocate of bioregionalism, acknowledges that tribal groups often engaged in war, but he stresses that this usually stemmed from resource pressures—pressures that he believes would not obtain in an ecologically balanced bioregional community (1985:123). Sale (1985:131) further claims that if the population of any one bioregion grows too large for its own resource base, it could either send out colonists to new areas or subdivide into smaller communities occupying ever smaller territories. Given general demographic expansion, the former strategy would work only for limited periods, while the latter would fail at the very start (subdividing resources would not make them any more abundant). Ultimately, the bioregionalists must fall back on the unsupportable proposition that by conforming to nature's ways, human communities will automatically achieve population stability and discover social harmony.

To attempt to force human communities at any scale into the confines of naturally constituted regions would require the fortitude of Procrustes. Rather than coinciding with either vegetational or physiographic units, human polities have typically transcended all manner of natural divisions. In so doing they may derive a kind of ecological integrity, but one based on complementarity rather than uniformity. A French geographic tradition, stemming ultimately from the work of Paul Vidal de la Blache, has long held that a political community is strengthened by its ability to draw on the diverse products of disparate environments (see, most recently, Braudel 1988). This notion, I would suggest, has far more to recommend it than does bioregionalism.

The environmental determinism that lurks behind radical environmentalism is nowhere more evident than in the murky philosophy of bioregionalism. According to environmental determinists, human cultures are ultimately shaped by the lands and climates in which they evolve. People living in cold climes are supposed to be more vigorous than those suffering oppressive heat, for instance, while desert landscapes, with their vast vistas, have been said to lead human minds inexorably toward monotheism (Huntington 1915; Semple 1911). Such simplistic thinking was enthusiastically embraced by early twentieth-century American geographers—an intellectual disaster from which the discipline has never fully recovered. When careful studies demolished its empirical basis and critical reflection revealed its close association with racism, environmental determinism was renounced by virtually every member of the academic community.

Yet bioregionalists would have to go back to this repudiated philosophy, albeit in a novel, *normative* form. While denying that human cultures are presently molded by their landscapes, bioregional advocates argue that they *should* be. In fact, they come very close to implying that it is our very escape from environmental conditioning that has led to the current ecological crisis. Geographer Stephen Frenkel (n.d.) has convincingly drawn out the parallels between modern bioregionalism and early twentieth-century environmental determinism, showing that adherents of both schools envisage the same fundamental associations between culture, human character, and politics on the one hand, and the natural environment on the other. As was true of the earlier movement, so with the present one; determinism leads in the end only to parochial and distorted views of both nature and culture. Moreover, notions of human freedom seldom flourish once deterministic philosophies are embraced.

Freedom and Centralization
Yet radical environmentalists' ultimate fear of large-scale organization lies precisely in its supposed incompatibility with local autonomy and individual freedom. To the eco-extremist, as scale increases so does central control, leading inevitably to the brutal efficiency of totalitarianism. But as with so many other green doctrines, supporting evidence is scarce indeed.

No one familiar with American legal history can seriously argue that our freedoms have diminished over the past several hundred years. The long-term trend has rather been one of a great expansion of personal choices. Freedom of ideas, freedom from gender and racial discrimination, freedom to challenge the government in the courts, and freedom to obtain state information have all been significantly enlarged. The 1980s clearly showed that countervailing tendencies exist, and continued struggle will certainly be necessary if we are to expand and even maintain our present scope of liberty. But it is undeniable that within the industrial democracies human freedoms have grown as economic and political scale has expanded. Eco-radicals may detect big brother peering over every technical innovation, but the capacity for resistance to state repression has been greatly enhanced as computers and copiers have spread and as telecommunications have grown more sophisticated. True, a totalitarian state can use technology against its citizens, but the people can also turn the tables. As Gary Miles (1990:648) demonstrates, "The ancient means of mass communication . . . served the interest of traditional hierarchical authority"; the modern means, in contrast, often subvert tyranny. It was for good reason that the Soviet Union under Brezhnev strictly limited the availability of copiers and microcomputers, and that fax machines are still illegal on the Gaza Strip.

Conclusion
The eco-radical vision of a small-scale social order in which personal ties prevail over bureaucratic structures provides a comforting antidote to the stresses of modern life. Virtually everyone longs, at some time, to escape the cold anonymity of the big city or the large firm. The problems identified here by eco-radicals are not all imaginary, and it would be foolish to embrace without reservation the impersonal and machine-like norms so often associated with large-scale social and economic formations.

But the best alternative, I would argue, is to struggle continually to *humanize* existing and emerging institutions. We can no more invent a small-scale, ecologically benign, bioregional future than we can return to an equally idealized primal past. Working within the imperfect forms of modern-day society is, admittedly, an anti-utopian gesture. We might improve society, but we will surely never perfect it. Those who are satisfied only with a vision that offers ultimate perfection will find little comfort in the liberal-moderate agenda advocated in this work.

The bioregional future imagined by eco-radicals would never provide them with the utopia they crave, even if it could be constructed exactly as envisaged. But that is the least of their problems. More pressing is the question of how we could *begin* to make the transformations necessary for the creation of the desired small-scale social world. As I have attempted to show, virtually any step taken in that direction would lead only to increased environmental degradation. It is one thing to dream of a pastoral America without cities, but quite another to imagine the process of deurbanizing the United States. Almost everywhere one looks, the eco-radical prescriptions are in the end just as harmful as the disease originally diagnosed.

 4

Technophobia and

Its Discontents

■ The Radical Position

Appropriate Technology

The third foundation of environmental radicalism is the belief that technological advance bears much, if not most, of the responsibility for the ecological crisis. Yet only the most extreme eco-radicals oppose all forms of modern technology. Far more actually favor technological developments, but only those that they consider appropriate. How one defines "appropriate" is, however, a matter of some debate. Evidently, what is appropriate to one writer can be considered deadly by another (Dobson 1990:98–101). Barry Commoner, for example, a vocal critic of many forms of modern technology, has been accused by another writer of evincing a "remarkable technological optimism" (Rubin 1989:45). Nonetheless, common to the entire movement is the thesis that technology has spun out of control and must be socially reharnessed if we are to avoid both the nightmarish future of a brave new world and the apocalypse of ecological meltdown.

Most green radicals define technologies as appropriate if they are small of scale, lend themselves to decentralization, emit little pollution, and do not require extensive consumption of natural resources. While the more sophisticated thinkers (for example, Porritt 1985) argue that some forms of high technology may be allowable, the larger eco-radical agenda makes it clear that appropriate technology must exclude virtually every innovation made over the past century—if not the past five millennia. If one were to take bioregionalism seriously and forbid trade across bioregional boundaries, virtually all metallurgy would cease. Of course, even the most strident may allow a few exceptions to the principle of autarky, but the denunciation of all truly complex technologies remains a staple of eco-radical thought. The editors of *Earth Island Journal*, for example, tell us that to save the earth, people should have their power lines disconnected, unplug their television sets, bury their cars, avoid all products that run on batteries, never travel on airplanes, and, most importantly, consume only products produced within their own bioregions (G. Smith 1990). In a "Neo-Luddite Manifesto," Chellis Glendinning (1990b:52) similarly advocates "dismantling the following destructive technologies": nuclear, chemical, video, electromagnetic, and computer. And in regard to agriculture, the eco-philosopher and small-scale farmer Wendell Berry (1977:212) informs us that the Amish are the "truest geniuses of technology."

The opposition to computers is perhaps the most controversial item on the antitechnology manifesto. Many green extremists, after all, use computers extensively in their own writing. To the uncompromising true believer, however, computers "cause disease and death in their manufacture, enhance centralized political control, and remove people from direct experience of life" (Glendinning 1990b:52). Berry (1990:170) provides another reason why the environmentally principled person should disdain computers: "I disbelieve, and therefore strongly resent, the assertion that I or anybody else could write better or more easily with a computer than with a pencil." Remarkably, Berry goes on to boast that his wife cheerfully prepares all of his manuscripts on a 1956 typewriter.

Underlying the eco-radical fear of technology is a longing for the preindustrial world of craft production, especially as typified by the medieval and early modern guilds of Europe (for example, Bookchin 1989:87; see Mumford 1966:272–73 for an early eco-romanticization of the guilds). The craft system, most green extremists assert, is both socially and environmentally superior to technologically oriented mass production. Craft industries are said to rely on natural, nonpolluting raw materials and to provide workers a humane environment in which to perform a variety of pleasant tasks. Consumers too would benefit under a craft regime, since they would be able to purchase goods that are far more durable and meaningful than the shoddy products of the industrial system. If select means of ingenious production developed by modern tinkerers could be harnessed to traditional, small-scale technology, craft manufacture would become more productive while remaining ecologically benign and socially beneficial.

The eco-radical critique of technology spans a wide range of products, many of which are not normally considered high tech. Even a simple material like cotton cloth, if woven by intricate machines that cannot be locally produced and maintained, implicates an unnecessarily sophisticated technological structure. Most fervent greens regard machine-driven production as objectionable in itself, since it allegedly alienates human beings from the creative process. As such, the eco-radical assault on technology can be seen as the spearpoint of a much broader attack on industrial production.

The anti-industrialism of radical environmentalism represents the survival of an old strain of extremely conservative thinking. One of the staunchest opponents of the English industrial revolution, the archreactionary Thomas Carlyle, also wished to return to the small-scale world of craft production—a system in which master was master and all workers had a secure place within a communal, if stratified, social order (see Williams 1983:71–86). While nominally opposing social hierarchy in all forms, contemporary eco-radicals echo Carlyle's basic thesis. And many have convinced themselves that preindustrial European society was not in fact highly stratified, at least when contrasted with regimes that were to follow. Glendinning, for example, argues that the Luddites, early saboteurs of factory machinery (and heroes of the Earth First!ers), "favored the old, relatively grass-roots economy over the *more hierarchical*, expansionist industrial capitalism" (1990a:180, emphasis added). Similarly, Brian Tokar, it may be recalled, saw medieval peasants as sufficiently "free of the pressures to overproduce" that they could devote most of their efforts to celebrations (1987:11).

In sum, the eco-radical critique of advanced technology, and of the manufacturing systems that accompany it, centers on four objections: it is dehumanizing; it is harmful to human health; it destroys the environment; and it entails an unabashed human arrogance toward nature. Each of these objections requires careful consideration.

Dehumanization

Following Jacques Ellul (1964) and Lewis Mumford (1966), most eco-radicals argue that the dehumanizing qualities of modern technology are most clearly evident in the labor organization required by industrial production. Factory work entails repetitive, unnatural tasks that are mind numbing if not brain destroying. Industrialism destroys the organic, life-affirming world of craftwork and replaces it with a crudely powerful but utterly lifeless production regime. Immured in industrial processes, workers begin to employ mechanistic metaphors for society and nature, thus contaminating and dehumanizing their very worldviews as well. Increasingly, people come to separate their long hours of drudgery from their *real* lives, their labor becoming nothing but a means to other ends. In contrast, in all preindustrial ages the modest amount of work people actually had to perform was thoroughly integrated with their basic life processes (Mishan 1973:71).

Eco-radicals also condemn industrial processes for destroying preexisting social relations. Before the industrial revolution, workers were comfortably supported by their families and natal communities; afterward they were immersed in the cold, cruel world of the machine, an environment that "squelches the individuality and uniqueness that fed the human spirit in times past" (Glendinning 1990a:144). Tender familial relations were replaced by rigid social hierarchies, with each worker becoming a slave to the masters of the mechanism. Here the argument against technological advance merges with those leveled against centralization and the development of capitalism, three processes that are pictured as conspiring to destroy humanity and nature.

Moderate as well as radical environmentalists often fear that technological developments will destroy jobs, thus threatening the populace with the dehumanization that accompanies mass unemployment (Dobson 1990:86; Young 1990:162, 200; Tokar 1987:87; Paehlke 1989:224). This is pictured as a process of long standing. In the early industrial revolution, spinners and hand-loom weavers lost their livelihoods to power machines, just as factory workers today sacrifice their jobs to process automation and robots. Newly redundant workers, in turn, have little option but to take even more menial and lower paying jobs in the service sector. As production becomes increasingly automated, eco-radicals tell us, high-skilled, well-paying jobs will grow ever more scarce. Some fear that this will lead to a vicious, downward economic spiral: the growing horde of poorly paid service workers will have less money to spend, undercutting the foundation of our mass-consumption economy. While technological optimists promise that robots might someday free humans from repetitive, mind-dulling tasks, in the absence of other employment options, such freedom will prove but a cruel hoax.

Many eco-radicals believe that the consumer goods produced by the factory system in shoddy forms and obscene quantities are themselves dehumanizing. Few contribute to truly meaningful activities, and as a package they demean the human spirit by lulling us into a stupor of consumer greed. Eventually we come to believe that happiness derives from the accumulation of mere things. "Material life alone flourishes," Donald Worster tells us (1985:58), "and for the manipulated mass man that seems to be enough: an iron cage with all the amenities will do nicely in the absence of other possibilities." Our machines produce abundant playthings that we discard into ever accumulating piles of waste once we tire of them, a phenomenon visible not only in our cancerous land-fills, but also in the toy-chests of every upper- and middle-class family in the United States. Electronic goods destroy our spirits more directly by disseminating droning technological propaganda (Tokar 1987:94, 95). Many eco-radicals would like to outlaw television, and the smashing of TV sets is an occasional ritual at their rallies and celebrations.

As discussed in chapter three, eco-radicals' strongest fear may well be that technological developments, particularly those in the fields of computers and telecommunications, will lead directly to increased central power, providing governments and corporations even more devious means of control. "In the past few decades," argues Brian Tokar (1987: 24), "the increasing computerization of all spheres of life has allowed methods of social control and surveillance to evolve to staggering proportions." According to Glendinning (1990a:140), even the telephone was "consciously developed to enhance systems of centralized political power."

Technology and Human Health

Environmental radicals also view modern technology as a direct threat to human health. Factories have always been dangerous places to work, and industrial accidents are still appallingly common. But far more deadly, they warn, are the modern chemical and nuclear technologies that attack the very substance of life. Within the high-tech factory, workers breathe a wicked fog of cancer-causing and immune-system-destroying substances. Many of these same death-dealing chemicals are disseminated throughout the biosphere; some are intentionally sprayed on crops, others are dumped in toxic waste pits from which they invariable seep out to contaminate the groundwater. Consumer products can also be toxic in their own right; the side effects of many modern medicines are more dangerous than the maladies they were designed to control, and even the most common plastics spew out small but potentially lethal quantities of formaldehyde and other unnatural gases. Unseen and subtle threats are ubiquitous in the modern world. Everything that generates an electromagnetic field, for example, presents a grave and immediate danger to all life forms. As our dwellings and offices grow ever more synthetic, increasing numbers of persons will develop the syndrome of "total environmental sensitivity"—in essence a debilitating set of allergies to the twentieth century.

Chellis Glendinning (1990a) offers the most concerted eco-radical assault on the medical hazards of modern technology. In her view, industrial society is suffering a virtual epidemic of cancers and immune-system disorders stemming directly from the poisons spewed into the environment by high-tech operations. Nature is now so wounded, she informs us, that it has some difficulty even supporting life. Glendinning's own social network is evidently composed largely of scarred technology survivors, individuals now organizing to challenge the central structures of high-tech society. Like most eco-radicals, they hope to reclaim a Luddite vision that will guide them in recreating a clean, safe, small-scale social world directly connected to the healing powers of nature. In such an intimate society, even such obnoxious commonplace contraptions as telephones will be unnecessary, since each person will be able to converse directly with everyone she or he knows (Glendinning 1990a:140).

Technology and Nature

While all eco-radicals decry the effects of technology and industrial production on human dignity and health, most fear primarily for nature. Technology's assault on the natural world is a fact of long standing; but in earlier days, when scale was small and techniques simple, damage was relatively minor and could easily be healed by nature's own recuperative powers. The more complex technology has grown, however, the further it has diverged from the basic processes of life and the more destructive each new advance becomes.

Barry Commoner (1990) argues that the creation of a major synthetic chemical industry epitomizes the decisive rift between nature and technology. In the post-Second World War era, factories began blindly to produce an ever increasing array of substances never before encountered. Many, perhaps most, of these chemicals turned out to be inherently destructive to life. Thus, even Commoner, who endorses a wide array of technologies deemed unacceptable by many other radical greens, calls for nothing less than the dismantling of the entire petrochemical industry. In an ecologically benign future world, he tells us, human society would rely on natural products that harmlessly biodegrade, like wood and cotton, and sedulously shun all unnatural plastics.

Although Commoner's denunciations of the petrochemical industry are powerful, Jeremy Rifkin (1983; 1989) again supplies a more thorough attack on modern technology of all varieties. Rifkin, who bases his argument on the second law of thermodynamics, tells us that the more we transform nature, the more quickly the universe's total energy will dissipate, leading ultimately to the remorseless state of "heat death." We would be wise, he cautions, to expend as little energy as possible so that we might forestall entropy's inexorable progress. Life may be doomed, but by dismantling our technological infrastructure we will be able to prolong its existence for a short spell.

An Affront to Nature

The eco-radical distrust of high technology has deeper roots than one might expect, roots extending well beyond fears about damage to organisms and ecosystems. Modern technology and its philosophical justifications are fundamentally viewed as arrogant *affronts* to nature even in the absence of firm evidence of actual harm. In essence, this is a secularized (or better, naturalized) version of the old religious creed that only God (nature) has the power to create, and that humans ought not to infringe on this divine prerogative. While few Christians now hold this view, it retains a certain currency in its eco-theological guise. The more we diverge from nature's patterns, green stalwarts believe, the more we deserve its wrath.

Two technologies are singled out as particularly offensive in this regard: nuclear engineering and gene splicing. Nuclear technologies, to be sure, are opposed primarily because of the very real dangers they present. Biotechnology, on the other hand, is denounced essentially because eco-radicals are wary of human beings playing God. The fear that mutant bacteria may escape from the laboratories and wreak havoc on the earth is real but secondary. Even in the absence of potential ecological hazards, most radical environmentalists would still find all forms of genetic engineering repulsive. To Bill McKibben (1989:166), such biotechnological manipulation represents nothing less than the "second end of nature." Jeremy Rifkin further claims that once we begin the process of manipulating genetic material we will not be able to stop: "As bio-engineering technology winds its way through the many passageways of life, stripping one living thing after another of its identity, replacing the original creations with technologically designed replicas, the world gradually becomes a lonelier place. From a world teeming with life . . . we descend to a world stocked with living gadgets and devices" (1983:252).

Another tenet of the eco-radical gospel is that technology further usurps God/nature in becoming a religious focus in itself. A belief in material progress is said to have emerged as the central creed of contemporary consumer society. Glendinning (1990a) argues that this myth is so deeply embedded in American culture that any technological advance is automatically hailed. Thus we blindly embrace every new development, no matter how lethal it may be. Indeed, technology drives madly forward virtually of its own accord; as Jacques Ellul (1964:83) explains, "[t]echnical activity automatically eliminates every non-technical activity." Rational decisionmaking and public debate, according to eco-radicals, have long since evaporated in the arid techno-worshipping atmosphere of modern society. This new secular religion may deny the reality of the apocalypse, but it actually functions to bring it about.

Eco-radicals also criticize the central tenet of technology worship, that progress must continue at all costs, as being in direct contradiction to the laws of nature. In nature's ecosystems, equilibrium prevails. If we are to coexist with the planet's other creations we must learn again how to fit within the earth's own modes of operation. Eco-salvation thus demands an emphasis on being rather than doing, on stationary existence rather than progressive movement. Behaviors that progress-obsessed moderns might regard as slothful are thus revealed as truly exemplary. Environmental historian Donald Worster (1985:335), for example, longs for an America in which "people are wont to sit long hours doing nothing, earning nothing, going nowhere, on the banks of some river running through a spare, lean land."

Questions about Science

The eco-radical antipathy to technology often extends to science as well. Denounced not only as the progenitor of harmful technologies, the scientific worldview is implicated in the intellectual rift that has torn humanity away from nature. "The modern scientific project," Dobson (1990:198) informs us, "is held to be a universalizing project of reduction, fragmentation, and violent control." Scientists are often depicted as brazen reductionists who attack the unity of nature by carving it up into isolated bits that they can proceed arrogantly to manipulate for their own satisfaction. Even the science of ecology is often suspected of harboring an unduly mechanistic and insufficiently spiritual appreciation of the unity of nature (Merchant 1989:9).

Environmental radicals also disparage science for its emphasis on specialization, a charge leveled against virtually every profession (for example, Milbrath 1989:207). Thus Young (1990:86–87) argues that since scientists now work in "large hierarchically organized teams, in which there is an increasing division of labor," the field itself has become "inherently conservative" and therefore ecologically destructive (see also Glendinning 1990a:24). Devall (1988:48–49) blithely informs us that "experts on nature" have "killed their positive feelings of identification" with the natural world and that "[s]tudents in natural resource sciences and management . . . are much like the guards in Nazi death camps." In fostering an atmosphere in which only the expert is accorded respect, science, technology, and capitalism are equally to blame. One gets the feeling from the more extreme texts that a specialist is little more than a half person, a being who has abandoned the meaningful in order to engage for hire in some petty and ecologically destructive activity. The specialist is thus but a cog in the death-dealing mega-machine (see Mumford 1966:200–201), utterly disengaged from the oneness of humanity and nature.

■ The Promethean View

Dystopia of Craft Production

Since eco-radicals idealize craftwork and disparage industrial production, it is first necessary to examine the social relations and environmental impacts associated with manufacturing prior to the industrial revolution. An appropriate starting point is Europe's medieval guild system, which several writers have touted as exemplifying social and ecological harmony. If the guild system can be proved socially exploitative, an important element of the eco-radical attack on industrialism will be discredited.

Eco-radicals are correct in arguing that working conditions within the guilds were, on average, far more humane than those imposed on the first industrial laborers. But medieval guilds most certainly were not the caring, familial institutions pictured in eco-radical fantasies. Many were authoritarian, if paternal, organizations; apprentices and journeymen worked firmly under the fists of their masters, and not all graduated to the status of independent craftsmen. Moreover, in heavy proto-industrial crafts, like metalwork, labor was hardly safe, let alone pleasant.

The medieval system of craft production is revealed to be even more objectionable when examined within its social context. The medieval world that made small-scale, socially organized craft production possible was rigidly hierarchical. The vast majority of Europeans in this period were impoverished peasants unable to buy anything produced by the guilds. In fact, until the 1820s members of the working class in France typically purchased their clothing second hand; only with the introduction of modern manufacturing and retailing could they afford to buy new goods (Reddy 1984:96). In preindustrial times, Fernand Braudel reminds us, the poor "lived in a state of almost complete deprivation" (1981:283). Sturdy craft objects were destined for the elite: the landed aristocracy, the ecclesiastical hierarchy, and the small but rising bourgeoisie within the towns. The entire guild system was founded on an extraordinarily inequitable distribution of resources. This should not surprise us; even today, craft goods (as well as many "natural" products) are purchased primarily by the rich, the only group able to afford them.

It was precisely because medieval and early modern craft production was so inefficient that only the truly wealthy could afford more than an extremely meager store of material possessions (Braudel 1981). While one could argue that poverty was widespread because the aristocracy monopolized consumption, it must be realized that the elite constituted a minuscule fraction of the population (Braudel 1982:466–72). Moreover, even many medieval and early modern aristocrats were not as wealthy as we enjoy picturing them. In preindustrial Europe there was nothing at all oxymoronic in the phrase "impoverished noble"; some were even reduced to begging for living (Blum 1987:25).

The material deprivation of medieval Europeans was not founded on a spiritual appreciation of the world uncorrupted by base material desires, as some eco-radicals seem to believe. Quite the contrary, material goods were actually valued more highly, relative to human life, than they are in modern society. As Braudel (1990:553) writes: "In the thirteenth century, '30 meters of Flanders cloth sold at Marseille [reached] two to four times the price of a Saracen woman slave.' [Such a price] may leave us 'pondering the mentality of the age, the price set on human life, the extraordinary value placed on a length of drapery from the Netherlands, and the considerable profits to be made from it by producers and *négociants*.'"

In select preindustrial societies, to be sure, certain social classes accumulated great hordes of material wealth, and in a few favored societies, such as in the seventeenth-century Netherlands, prosperous middle classes grew to substantial proportions (Schama 1988). But such wealth as did exist was made possible only by large-scale transregional exchange or imperial plundering. In the immediate preindustrial period, much of Europe's prosperity rested on trade with, and exploitation of, the rest of the world. Even in the medieval period, trade networks spanned the subcontinent and extended ultimately to many far reaches of the globe. Bioregionalism was never an operative principle in the world of the guild.

One should also recognize that centuries before the mechanization of cotton spinning, Europe as a whole had been benefiting from technological innovations that many eco-radicals would disparage. Historian Jean Gimpel (1976) argues that the first industrial revolution occurred precisely in the Middle Ages. Medieval engineers and entrepreneurs were already damming rivers to harness water power, digging for coal in strip mines, and processing select raw materials in reasonably large-scale operations. Such technical advances vastly increased the subcontinent's meager store of wealth, but they also brought about a sometimes substantial level of industrial pollution. Gimpel's (1976:86) description of tannery wastes is apposite here: "Tanning polluted the river because it subjected the hides to a whole series of chemical operations requiring tannic acids and lime. Tawing used alum and oil. Dried blood, fat, surplus tissue, flesh impurities, and hair were continually washed away with the acids and the lime into the streams running through the cities. The waters flowing from the tanneries were certainly unpalatable, and there were tanneries in every medieval city."

In short, the preindustrial world was far from the ecological and social paradise imagined by some eco-radicals. Only by embracing an idealized and ultimately fraudulent picture of life before mechanization can one accept the eco-radical faith in craft production.

Disease: Technological and Natural
The second prong of the eco-radical attack on modern technological products and processes lies in the assertion that they constitute a massive threat to human health. While there is certainly much truth in this general proposition, the more extreme writers go so far as to argue that health standards have been progressively declining as our environments have grown more synthetic. To disprove this strong version of the technophobic disease thesis, we can simply compare incidents of death and disease under preindustrial and industrial regimes.

No one acquainted with the rudiments of medical history could deny that health has vastly improved since the industrial revolution. Most of the credit for such amelioration belongs precisely to the medical, dietary, and sanitary advances associated with the transition to industrialism. One has only to examine average longevity, which stood in the United States at a miserable forty-seven years as recently as 1900, to grasp the magnitude of progress over this period. If we go back to medieval Europe, socio-ecological idyll of many eco-radicals, we find that in some villages average life spans were as low as seventeen to eighteen years (Cohen 1989:124).

By other indices as well, the health standards of most preindustrial regimes were atrocious. Again, consider medieval and early modern Europe. As Braudel (1981:91) relates, the ancien régime was characterized by "very high infant mortality, famine, chronic under-nourishment, and formidable epidemics." Moreover, nonelite Europeans were contaminated by a wide variety of toxins on a regular basis. Few even experienced the delights of breathing clean air, for the atmospheres of their own dwellings were horribly polluted. "It is difficult . . . to comprehend," writes Norman Pounds (1989:187), "how fetid and offensive must have been the air about most cottages and homes." Indeed, indoor air pollution has long been (as it perhaps still is) a greater contributor to respiratory illness than industrial airborne waste.

But the most severe toxic pollution problem of the premodern world was associated with natural poisons produced by molds infecting the food supply. "Everyone suffered from food that was tainted," Pounds reminds us, "and the number who died of food-poisoning must have been immense" (1989:213). Especially pronounced where rye was the staple food, poisons produced by the ergot and *Fusarium* molds massively suppressed immune systems, reduced fertility levels, brought on delusions and sometimes mass insanity, and reduced blood circulation to such an extent that gangrene in the lower extremities was commonplace (Matossian 1989).

Even where the food supply was safe, poor nutrition resulted in widespread immunological stress. Infectious diseases were rife, and periodic plagues would decimate most populations in a cruel manner. Water supplies, especially in towns, were so contaminated by human waste as to become deadly in their own right. Skin and venereal diseases were often rife and difficult, if not impossible, to cure. Other scourges abounded, including those—such as leprosy—that have been virtually eliminated by modern medicines and sanitary techniques. Individuals deformed by genetic inheritance or accident typically led short and brutal lives. And every time a woman went into labor she faced a very high risk of dying.

This cursory review of the horrors of preindustrial European life may seem a pointless exercise in overkill; all of this is, or at least used to be, common knowledge. But it is important to recall in detail the kind of social environment many eco-radicals would seek to recreate. And were we to adhere strictly to the tenets of bioregionalism, even the levels of prosperity achieved in the medieval world would be difficult if not impossible to maintain without first experiencing a truly *massive* human die-off.

If the eco-radical vision of the preindustrial past is highly distorted, its view of the past half century is hardly more realistic. The notion that the last fifty years have seen a cancer epidemic visited upon the world is based on highly questionable statistical evidence. Glendinning (1990a: 56), for example, argues from the fact that while only 3 percent of total deaths in the United States in 1900 could be attributed to cancer, now one in every three persons can expect to contract the disease. There are several problems here. First, it is misleading to compare death rates in the former period and disease rates in the latter. More importantly, while it would be foolish to deny that many cancers are linked to industrial toxins, the long-term rise in cancer cases at least partly reflects the fact that in 1900 most persons died of other causes before they had the chance to experience this disease. As more individuals survive into old age, the total number of cancer patients will necessarily increase.

Eco-radicals are absolutely right to decry the appalling illnesses induced by toxic wastes and agricultural chemicals. Promethean environmentalists readily join in the struggle first to reduce, and ultimately to eliminate, human-generated environmental toxins. But the elimination of toxic waste is a technical problem that demands technical solutions. Dismantling all modern industry is not the answer. Moreover, we must carefully balance concerns about human health with other environmental issues, many of which are more pressing. As *The Economist* reports (March 30, 1991, p. 28), the Environmental Protection Agency (EPA) has at times "behaved more like a cancer-prevention than an environmental agency." To spend billions of dollars attempting to reduce slightly our cancer risks, while entire ecosystems perish for the want of a few hundred thousand dollars, shows a questionable and highly anthropocentric sense of priorities.

Industrial Amelioration

The eco-radicals' critique of industrial labor conditions proves to be more instructive than their disease thesis. Early factories did subject their workers to a brutal existence, and many plants operating today in the less-developed parts of the world are equally dismal. In fact, much evidence suggests that the industrial revolution was accompanied by a general downturn of average living conditions, a decline that lasted several generations (Braudel 1984:614). Both wage levels and industrial work conditions were eventually to experience steady improvement, but progress remains unacceptably halting. Moreover, certain industries, such as meat packing in the United States, sometimes experience sharp regressions toward increased workplace danger. And even where they are safe, assembly-line jobs remain mind-dullingly tedious and poorly remunerated. Eco-radicals are also correct in pointing out that industrial toxins present a massive threat to worker health, and that the lack of attention given to this problem has been shameful.

But despite these massive remaining problems, the solutions to workplace brutality and injustice do not lie in stepping backward to a system of craft production. Rather, we should continue to move forward, through both technological improvement and regulatory advance. Together, these two forces have been responsible for a tremendous amelioration of industrial conditions. The vigilant enforcement and continued extension of safety regulations remains absolutely essential. But equally important is the continuation of the automation process itself. In an eco-Promethean future, hazardous and deadening tasks would not be performed by human beings at all, but rather by unfeeling robots and other automatic devices, many of which may well exist at the nano scale of molecular assemblages.

Much improvement has already been brought about by automation. This is readily visible if we compare, circa 1990, the relatively safe and humane industrial work conditions found in western Germany with those located in technologically stagnant eastern Germany. Eastern Germans working in the glass industry, for example, suffered horribly in: "investment-starved factories [that] are hazardous industrial relicts. Jenar's glass-makers toil in temperatures that can reach as high as 140 F. They blow molten glass by hand a few feet from fire-belching open hearths" (*Fortune*, April 22, 1991, p. 224).

Yet even where automation clearly reduces workplace hazards, eco-radicals still reject it for generating massive unemployment. Although automation does result in an initial displacement of workers, an issue that can and should be humanely addressed through retraining programs, in a healthy industrialized economy its long-run consequences are clearly positive. Substituting capital for labor, if done intelligently, boosts productivity, creating a larger economic pie for society as a whole. Moreover, as Joel Mokyr (1990) explains, technological progress itself is a positive-sum game, one in which winners far outnumber losers. Increased productivity leads to economic expansion, with the end result being that extinguished jobs will be replaced, on aggregate, by *better* paying jobs in other sectors. How else can one explain the phenomenal rise in living standards that eventually accompanies successful mass industrialization? Alternatively, how can one account for the modest unemployment rate and mounting prosperity of Japan, the world's pacesetter in automated production? If the eco-radical unemployment thesis had any merit, industrialization would have been a self-canceling process from the very beginning.

Groups directly threatened by technological advance have always considered it socially destructive, and many have been able to prevent or delay the introduction of highly beneficial innovations. Few eco-radicals, I imagine, would today prefer roman to arabic numerals, and only the most extreme would object to printing. Yet as Mokyr (1990:179) relates, in 1299 the bankers in Florence were forbidden the use of arabic numerals, whereas "in the fifteenth century, the scribes guild of Paris succeeded in delaying the introduction of printing into Paris by 20 years." The modern opposition to computers and many other forms of technology demonstrates the same kind of thinking that led Parisian scribblers to resist Gutenberg's invention.

Those who believe that mechanization brings massive unemployment overinterpret the high rates of joblessness experienced in the United States (and much of Europe) in recent years. Several factors conspired to generate widespread unemployment in this period. One of the more important ones was demographic; when the huge baby-boom generation began to seek jobs, employment opportunities naturally diminished. The same era also witnessed the massive entrance of women into the job market, as well as a continuing influx of job-seeking immigrants, both unskilled and skilled. But the most significant reason for joblessness—as well as for America's general economic malaise—was simply the productivity slowdown. As both Lester Thurow (1985) and Paul Krugman (1990) clearly illustrate, lagging American economic productivity, and our consequent failure to remain internationally competitive, is the root cause for economic alarm. To a significant extent, the productivity crisis reflects a lack of capital investment, itself a symptom of the short-term thinking characteristic of many American executives, rather than the overinvestment feared by eco-radicals.

Japan offers a perfect counterpoint to the eco-radical unemployment thesis. A fierce labor shortage has encouraged Japanese managers to invest heavily in automation and robotization. Although often not initially profitable, long-term automation projects have been made possible by the strategic autonomy of Japanese corporations. American firms, by contrast, are usually required by Wall Street to seek extremely short payback periods for their investments. Indeed, this is a major reason why the Japanese economy continues to rack up much more impressive productivity gains than does the American economy. Not surprisingly, Japan has in the process come to dominate the global robotics industry. American robotics firms, once the global pioneers, are now virtually extinct (*Fortune*, April 16, 1990, pp. 148–53). As the Japanese economy continues to grow and as Japanese firms continue selectively to replace labor with capital, resources are made available to retrain workers for more skilled and higher paying employment elsewhere in the Japanese economy.

Radical environmentalists not only misrepresent the relationship between automation and unemployment, but they also cling to an out-dated vision of the former. Successful automation no longer necessarily entails mass production. On the contrary, through flexible automation, small, individualized batches of goods can now easily be produced. Flexible automation has the potential to fuse the best features of the old and the new production regimes; according to one optimistic prognosis, "the craft-era tradition of custom-tailoring of products to the needs and tastes of individual consumers will be combined with the power, precision, and economy of modern production technology" (Dertouzos et al. 1989:131). Where radical greens see only increasing uniformity, those who are actually observing the evolution of technology discern rather uniformity's demise.

Toxin Production and Destruction

The eco-radical critique of technology becomes most vehement on the subject of toxic by-products and other pollutants. Although concerns here are absolutely on target, the solutions proposed are fundamentally misguided. Rather than dismantling our technological infrastructure, a politically unfeasible agenda to say the least, we should reengineer it so that destructive contaminants never reach the environment in the first place. Developing clean production systems will require sustained technological advance, as well as a tremendous rechanneling of capital, but it is by no means beyond the reach of human ingenuity. Political timidity and short-term economic fixations, not technology per se, dictate that we continue to contaminate our environment with deadly substances.

For the sake of brevity, the following section concentrates on the more dangerous toxic discharges rather than the more common forms of air and water pollution. The former are both more debilitating in the long run and more challenging to eliminate. If we can continue to develop our technological apparatus while eliminating toxic wastes, we should easily be able to handle the more conventional varieties of pollution.

Most toxic wastes are composed of chemical compounds that can be reduced, with some effort, to their harmless constituent elements, such as carbon and hydrogen. Various means of decomposing toxins are currently in use or being developed. Some rely on physical processes, particularly focused solar energy or combustion at high temperatures, but many of the more sophisticated techniques employ biological metabolism. Certain species of bacteria thrive on, and devour, many varieties of toxic sludge. By providing these microorganisms with a favorable environment, decomposition may be greatly accelerated. And if genetic engineering fulfills its promise, crud-devouring bacteria may be expected to work much more efficiently in the future (see Kokoszka and Flood 1989; National Research Council 1989; Johnston and Robinson 1984; Omenn and Hollaender 1984).

In many instances firms may find it more efficient to use rather than to destroy (or dump) what were formerly waste materials (Freeman 1990). Recovery of wastes has in fact been on-going for several hundred years (Mowery and Rosenberg 1989:55; Wilkinson 1988:95). Capturing such materials and recycling them in other industrial processes reduces their contact with the environment. Further chemical processing, moreover, can render certain kinds of wastes both inert and useful. Several companies have already learned to profit from what were until recently polluting by-products; Du Pont, for example, has discovered that it can sell its acid iron salts to wastewater treatment plants (*Fortune*, February 12, 1990, p. 48). One company's garbage is often another's raw material, and the interfirm marketing of waste products is a growing business (Patterson 1989).

Those forms of toxic waste containing heavy metals, which are dangerous in their elemental states, are obviously inappropriate candidates for decomposition. But genetically engineered microorganisms can again be employed, in this case to collect metallic molecules so that they can more easily be sequestered (Higgens 1985:235). Advances in membrane technology and filtration systems will also allow more efficient isolation of heavy metals as well as other forms of toxic waste. Once collected, lead, mercury, and other metals can be recycled, offering financial benefits as well as reducing the need for further environmental disruption through mining.

As companies learn to reduce their waste streams through these and other methods, they sometimes discover that their operations grow more efficient in the process. Minnesota Mining and Manufacturing, for example, through its Pollution Prevention Pays program, has realized savings that already amount to over one billion dollars (*Fortune*, February 12, 1990, p. 48). As recycling, decomposition, and sequestering techniques grow more sophisticated, increasing numbers of companies can be expected to adopt them—especially if regulations grow more stringent. As this occurs, economies of scale will emerge, further reducing the costs and increasing the economic benefits of pollution control in a virtuous spiral of environmental cleansing.

Synthetic Materials: Sin or Salvation?
Eco-radicals disdain synthetic materials (typified by plastics) in part because they are not biodegradable. From the Promethean perspective,

however, resistance to rot can be highly advantageous. Nonbiodegradable materials are, on aggregate, easier to recycle than are their natural alternatives. Paper fibers, for example, break down during the recycling process, limiting their potential for reuse. Paper also spoils if improperly stored, rendering it unsuitable as a raw material. Many synthetics, by contrast, can be recycled a vast number of times.

Admittedly, plastic beverage containers cannot be sterilized, severely restricting their potential for immediate reuse. They can, however, be melted and reextruded to form durable products, a process known as secondary recycling (Leidner 1981). Recycled plastic has long been a substandard product, limiting its applications, but researchers at Battelle Memorial Institute have recently developed a process by which plastics can be reextruded with no decline in quality (*Business Week*, April 15, 1991, p. 72). While glass, also a nonbiodegradable product, remains an environmentally superior food and beverage container, plastic is clearly preferable to wood (see below) in the manufacture of consumer durables. Not surprisingly, Japan and Western Europe have pioneered the development of secondary recycling techniques. Such processes have long been unappreciated in the United States, in part because of our heavy subsidization of the wood products industry (Leidner 1981:157).

Contrary to eco-radical doctrine, biodegradation itself, given our current waste disposal system, can generate serious environmental contamination. Most paper bags in the United States, for example, are disposed in sanitary landfills. Once buried they are isolated from oxygen, and thus decompose, if at all, anaerobically. Anaerobic decomposition, in turn, produces methane, a very powerful greenhouse gas. Plastic bags, in contrast, are relatively inert, limiting rather than magnifying their environmental damage. Only in areas where garbage might find its way into an aquatic environment should plastic be avoided as intrinsically damaging.

The outgassing of potentially harmful molecules by plastics and other synthetics is a threat that must be taken seriously. Since demands for energy efficiency will lead to the construction of increasingly airtight buildings, indoor pollution will become a mounting hazard in the absence of concerted action. We may expect technological advances in organic chemistry, however, as well as advances in ventilation and filtration systems, to reduce the problem.

Promethean environmentalists agree with Arcadians that some technological products are intrinsically destructive. A prime example would be the ozone-attacking chlorofluorocarbons (CFCs). Given the proper incentives, however, engineers are usually able to devise relatively benign substitutes in a remarkably short time, as the CFC dilemma has already demonstrated. Devising such substitutes, however, requires a major commitment of economic resources and especially scientific expertise. But scientific expertise is itself under attack by the eco-radical community.

Science, Monitoring, and the Environment

The eco-radical attack on the reductionism and specialization inherent in science is environmentally threatening in its own right. If we were to abandon scientific methodology we would have to relinquish our hopes that environmentally benign technologies might be developed. Advances in solar power will not come about through holistic inquiries into the meaning of nature.

The scientific method also must be applied in environmental monitoring. Had it not been for highly specialized measuring techniques, we would not have known about the CFC threat until it was too late. Moreover, the requisite devices would never have been made were it not for the organization of the scientific community into distinct specialties, each framing its inquiries in a reductionistic manner. To avoid environmental catastrophe we need as much *specific* knowledge of environmental processes as possible, although it is also true that we must improve our abilities to combine insights derived from separate specialities. Much greater emphasis must be placed on basic environmental science, in both its reductive and synthetic forms, a project that would be greatly hindered if we insist that only vague and spiritually oriented forms of holistic analysis are appropriate.

Eco-radicals can be expected to counter that environmental monitoring is only necessary in the first place because of industrial poisoning; dismantle industry, and environmental science will cease to be useful. Although seemingly cogent, this argument fails on historical grounds. As discussed previously, toxins can be produced by nature as well as by humanity. For centuries Europeans attributed the delusions they suffered after eating ergot-infected bread to evil spirits. Thousands of women were burned at the stake because of the fearful reactions of a patriarchal, religiously fundamentalist society to the psychological effects of an unknown, natural, environmental toxin. Once scientists, using *specialized* techniques, isolated the agent, ergotism and its associated social pathologies began to disappear (Matossian 1989).

In many different fields specialized scientific techniques are now proving invaluable for the efforts to control pollution and preserve natural diversity. For example, the development of biosensors—mechanisms that "combine biological membranes or cells with microelectronic sensors"

(Elkington and Shopley 1988:14)—promises vastly improved means of pollution detection. Similarly, the development of Geographic Information Systems (GIS), based on the construction of spatialized computer data bases, has allowed geographers and planners to predict the ecological consequences of specific human activities and thus minimize deleterious impacts on critical ecosystems. Nature Conservancy field agents, for example, have found GIS a useful tool in devising conservation strategies for Ohio's Big Darby Creek, one of the Midwest's few remaining clear-flowing streams (Allan 1991). Geographers have also repeatedly proved the utility of satellite image interpretation for developing and implementing conservation plans at the national level (Elkington and Shopley 1988). We may expect eco-extremists to have little patience with such philosophically impure forms of environmental work. Yet rejecting such techniques outright would only intensify environmental destruction.

Natural Products and the Destruction of Nature

Assessing the eco-radical aversion to technology also requires considering the environmental effects of natural, low-tech products. Although this is an extremely intricate issue, many natural substances actually prove to be far more ecologically destructive than their synthetic substitutes.

Wood provides a good example of a destructive natural product. By relying on wood for building materials, simple chemicals, and fuel, countless societies have deforested their environments. The switch from wood to coal as an energy source helped save European forests from total destruction in the early modern age, just as it did for American forests in the 1880s (Perlin 1989). Pressures on forests were also reduced when the Leblanc process was developed, allowing soda to be manufactured from salt rather than from woodash. (This discovery also drastically reduced the cost of soap, tremendously benefiting human health.) The Leblanc process was, however, highly polluting, but the subsequently developed ammonia process proved to be considerably cleaner and more efficient as well (Mokyr 1990:121).

The common belief that wood is an environmentally benign and renewable resource is dangerously naive. Forests are effectively renewable only where population densities are extremely low. Unfortunately, areas of requisite density are becoming increasingly rare throughout most of the world. In the contemporary Third World, technological deprivation forces multitudes to continue living within an unsustainable wood economy. Poor women often spend hours each day scrounging for firewood, a process both ecologically and socially destructive. Where electricity is available and affordable—as it should be everywhere—deforestation rates decline drastically.

The use of wood as a construction material in contemporary industrial societies is also environmentally devastating. The havoc wreaked on Southeast Asian tropical rainforests by the Japanese construction industry is a commonly acknowledged environmental outrage (see Laarman 1988), but the effect of American house-building on our own temperate rainforests is hardly less objectionable. Economic considerations ensure that even sustainably and selectively harvested forests are degraded as wildlife habitat. Foresters shudder at the idea of preserving dead and dying stumps that might form disease reservoirs, but it is precisely such hollow trees that provide denning sites for many mammals and nesting sites for many birds. While radical environmentalists might argue that we should therefore adopt less efficient forms of forestry, the problems that would ensue because of the resulting decline in timber yield are not addressed. With a growing population continuing to demand lumber, a deintensified forest industry would be forced to seek new supplies elsewhere, thus degrading even larger expanses of land. In the end, only by developing substitutes for wood can we begin to create an environmentally benign construction industry.

Many wood substitutes are readily available. Concrete, for example, is easily and efficiently employed in all manner of construction. Yet eco-radicals like Catton (1980:135) warn against using concrete on the grounds that it is a nonrenewable resource. I would counter that the prospect of abandoning cement making and aggregate mining for fear that we will exhaust the planet's supply of limestone, sand, and gravel is an example of green lunacy. We might as well dismantle the ceramics industry for fear of exhausting the earth's clay deposits.

Paper, another natural product, embodies extraordinary environmental destruction. Papermaking remains one of the most polluting industrial processes in existence. Even if paper-mill wastes can be minimized (at some cost), and even if recycling becomes commonplace, paper production will continue to demand vast quantities of wood. Resource economics dictate that the necessary quantities of fresh pulp be derived largely from small, fast-growing trees, generally harvested in clear-cuts. The resulting pulp plantations are typically as ecologically impoverished as agricultural fields. By continuing to prefer paper to synthetic and electronic substitutes, we only ensure the needless degradation of vast tracks of land.

Many other examples of the ecological destruction inflicted by natural products could easily be cited. The damage entailed in cotton production, for example, was noted twenty years ago by Ehrlich (cited in Paehlke 1989:60). While cotton could be cultivated without biocides, yields would plummet, necessitating a substantial increase in acreage to meet the present demand. The area devoted to cotton is expanding at a rapid pace already, due both to population growth and to the mounting demand for natural fibers. Vast expanses of natural vegetation are now being cleared in order to grow cotton and to supply it with the water it requires. To provide high-class textiles, the Ogallala aquifer of America's southern Great Plains is being depleted, rain forests in Central America are being devastated, and the extensive Sudd Swamp of the southern Sudan is being threatened with drainage.

The standard environmentalist credo that renewable resources are intrinsically superior to nonrenewables rests on two fundamental errors. First, both eco-radicals and old-fashioned conservationists presume life to be so abundant that through wise use, directed either by primal affinity or scientific management, humans can obtain their needs organically without detracting from other species. Second, both camps have assumed that nonrenewables are so scarce that if we dare use them they will be quickly exhausted. Both principles are suspect.

In fact, the primary organic productivity of the planet is essentially limited. The more living resources are channeled into human communities, the more nature itself is diminished. The essential nonrenewable resources, by contrast—elements such as silicon, iron, aluminum, and carbon—may be tapped in extraordinary quantities without substantially detracting from living ecosystems. Aluminum and silicon are so wildly abundant that it is ludicrous to fear that we will exhaust the earth's supply. Moreover, except in nuclear processes, elements are never actually destroyed; as recycling and sequestering techniques are perfected, resource exhaustion will become increasingly unproblematic. Even coal and oil would be fantastically abundant if only we would cease the insane practice of burning them and instead, as suggested by Amory Lovins, dedicate the remaining supplies to the production of synthetic organic materials (see Paehlke 1989:77).

A society based on the principles of Promethean environmentalism will cease as much as possible to provision itself through the killing of living beings, be they animal or plant. Instead, it will strive to rely on nonliving resources, whether formed of long-dead matter, like oil and coal, or simple inorganic substances, like silicon. Learning to build our material world out of nonliving resources will entail both high-tech and low-tech methods. Simple technologies using stone, brick, tile, and concrete have eventually been devised by all forest-destroying civilizations

(Perlin 1989), and they continue to be useful. More sophisticated approaches entail the development of superior composite materials and synthetic organic compounds. Many such products deliver additional environmental payoffs; certain composites, for example, are both strong and light, giving them profound advantages for energy-efficient transport systems.[1]

Telecommunications and computer systems present another field in which technological advance could yield vast environmental benefits. Consider the advantages of electronic mail (E-mail) over the conventional mail delivery system. To operate the latter, entire forests must be dedicated to paper production, while huge fleets of trucks and airplanes must be maintained and fueled for parcel delivery. Transmission of E-mail, on the other hand, requires only silicon chips, glass cables, and energy-sparing pulses of information. Similarly, one would hope that improved transmission of video images will eventually obviate the need for much—perhaps most—business travel. The sooner we embrace the telecommunications revolution and dispense as much as possible with paper and with unnecessary personal contact, the less environmental damage our communications will inflict.

Energy

As all environmentalists recognize, deriving the bulk of our energy from fossil fuels is an unsustainable practice. Oil, gas, and coal deposits will eventually be depleted, undermining in the process the future of the synthetic organic chemical industry. The combustion of fossil fuels is also intrinsically damaging to the environment, especially by releasing stored carbon that threatens the planet's heat balance.

Many environmentalists have proposed that we obtain energy by burning renewable resources. Biomass derived from agriculture and forestry, they claim, could be endlessly recreated in future crop cycles (Porritt 1985:177). But as the preceding pages have argued, large-scale biomass conversion would prove to be an ecological catastrophe. To supply our energy needs, tremendous expanses of natural habitat would have to be converted to croplands or tree plantations, resulting in a massive reduction of natural diversity.

The solution to the energy bind lies, as most members of the environmental community realize, in a combination of solar power and conservation. What eco-radicals fail to recognize, however, is that both effective conservation and the commercialization of solar energy demand highly sophisticated technologies. The modern frontiers of energy conservation may be found in such areas as low emissivity windows, energy-sparing fluorescent light bulbs, and computer-integrated sensor systems (Fickett et al. 1990; Bevington and Rosenfeld 1990). Due to a wide variety of such advances, the energy intensity of American industry in fact declined at a rate of 1.5–2 percent per year between 1971 and 1986, allowing industrial production to increase substantially while energy consumption actually fell (Ross and Steinmeyer 1990).

When it comes to harnessing solar power, technological achievements are even more vital. Admittedly, several important solar applications demand little technical sophistication. Simply by placing windows properly a significant power savings can be realized. But in order to do something slightly more complicated—such as heat water—certain high-tech applications are essential. The simplest passive solar water heating systems usually rely on components made of plastic, a substance many eco-radicals would like to ban.

But to address our needs for an ecologically benign power source, solar-generated electricity must be commercialized on a massive scale. No matter how this is done, significant technological advances will be necessary.

A certain amount of electricity can be indirectly obtained from the sun by harnessing wind energy. Careful estimates show that fifteen American states could supply all of their electricity needs from environmentally benign wind-driven turbines (Weinberg and Williams 1990). As incremental advances are made in turbine technology, wind power may be expected to become ever more competitive with conventionally obtained power. Such improvements are already being seen, the cost of wind-generated electricity having dropped nearly 90 percent since 1981 (Weinberg and Williams 1990).

Yet in California, the state most committed to this alternative energy source, eco-radicals have recently begun to struggle against wind power development. The reasons: high levels of bird mortality caused by the spinning blades (admittedly a serious problem), and the fact that wind farms are an unsightly affront against the pristine landscapes in which they are typically located (discussed in Paehlke 1989:99). That only a minuscule portion of the state even has the potential for wind power development has not lessened their outrage. Here again many eco-radicals demonstrate a highly dangerous opposition to an environmentally promising technology.

Although wind power may someday be crucial in meeting the energy needs of a few windy states, direct solar power is far more promising as a possible solution to the energy crisis. Several competing technologies, notably solar thermal and photovoltaic, may supply tremendous amounts of relatively cheap electricity in the near future (see Weinberg and Williams 1990). Of the two, photovoltaics, or PVS, show the most promise.

The cost of PV generated electricity has plummeted in recent years as solar cell efficiencies have increased and as economies of scale in manufacturing have begun to appear. At some 20 cents a kilowatt hour, PV electricity is now competitive with conventionally derived electricity in locations not yet connected to power grids. With continued investment in both design and manufacturing techniques, PV costs are expected to continue to fall, offering the possibility of an impending breakthrough into the mass market. One especially promising horizon in photovoltaics is the development of solar cells composed of thin film amorphous silicon, which may potentially prove both inexpensive and highly efficient. Manufacturers are also conducting research on nonsilicon materials, including copper indium diselenide, gallium arsenide, and cadmium telluride, all of which offer specific advantages. Arco Solar, for example, has recently reported a very impressive 15.6 efficiency rate using translucent silicon and CIS (copper inidium diselenide) (Bernstein n.d.:10; Ogden and Williams 1989). The most exciting recent breakthrough, however, is the development of silicon bead technology, pioneered by Texas Instruments and Arco Solar. This method of production appears to be so inexpensive that some researchers believe that it will soon make solar electricity fully competitive with conventional sources (*Business Week*, April 22, 1991, p. 90).

As large-scale PV generation becomes more feasible, the difficulties of storage will grow more prominent. Since PV electricity flows only when the sun shines, the challenge is to deliver power at night and on cloudy days. The lead-acid batteries now used for storage are both expensive and inefficient. Research is being conducted, however, on sodium-sulfur and zinc-bromine batteries that "store more energy in less space, offer longer lifetimes, and cost less than lead-acid batteries" (Bernstein n.d.:14). Superconducting magnetic energy storage may offer even greater benefits, but only if a daunting series of technical and economic obstacles are first overcome (Bernstein n.d.).

Although a variety of problems remain, the successful commercialization of photovoltaics, unlike fusion power, will not require major scientific breakthroughs. Continued incremental advances along several fronts can be expected to render PVS increasingly competitive with conventional electricity sources. Importantly, PVS offer greater potential for the realization of economies of scale than do most competing power sources because they are constructed in the factory rather than the field

(Ogden and Williams 1989:50). The difficulties currently being faced by the PV industry stem ultimately from its own immaturity—and from the negligible amount of governmental assistance that it has received—rather than from any intrinsic failings.

Yet even if solar-generated electricity were soon to fulfill its promise, the challenge of supplying energy for mobile applications would remain. Several automobile companies (most notably GM) have made great strides in designing electric cars, although the development of lightweight storage batteries remains a stubborn obstacle. Equally promising is the creation of hydrogen-powered vehicles. Unlike other fuels, hydrogen burns cleanly, releasing little but water vapor. In an integrated, environmentally benign energy system, solar-generated electricity could be used to reduce water to its constituent elements, supplying in the process high-energy hydrogen fuel (Weinberg and Williams 1990). Certainly many challenges remain, especially that of rendering hydrogen both safe and easily transportable. But several companies, notably Daimler-Benz, BMW, and Mazda are presently working on these problems (*Business Week*, March 4, 1991, p. 59; Ogden and Williams 1989).

Tragically, many eco-radicals have joined anti-environmentalists in disparaging the possibility of a transition to a full-fledged solar economy. Radicals voice a variety of predictable concerns. Many consider the devotion of large expanses of land to solar collectors completely unacceptable. Especially galling is the prospect of relatively pristine desert environments being sacrificed for energy collection. More fundamentally, eco-radicals shun photovoltaics because of the sophisticated technology required (see Dobson 1990:103)—the same technology implicated in the feared information revolution. PV manufacturing also generates toxic wastes, which many regard as reason enough to ban the entire industry. Moreover, PV systems could not possibly be constructed and maintained on a bioregional basis, thereby excluding them from the realm of the environmentally correct.

The anti-environmental opposition to solar power is a bit more curious. While anti-environmentalists exude unshakable optimism when considering ecologically destructive technologies such as nuclear fission, their forecasts quickly turn dismal when confronted with ecologically benign innovations. Dixy Lee Ray (1990:128), for example, dismisses solar power out of hand, stating simply that "solar generated electricity is not a practical alternative." If the prognosis for solar power were really so miserable, one might well wonder why the Japanese government and major Japanese corporations are pursuing it so avidly. According to the logic of Promethean environmentalism, solar technologies *can* provide our energy needs, but only if we are willing to adopt a long-range economic perspective. Seen in this light, the antisolar stance of writers like Ray seems little more than a pathetic attempt to justify the short-term thinking that is presently leading the American economy along a sustained curve of relative decline.

Techno-environmentalists like Oppenheimer and Boyle (1990) argue that if we have the foresight and fortitude to develop a solar-based economy, we can both avert the potential catastrophe of global heating and propel the United States into a renewed era of sustained economic growth (the so-called fifth wave of the Kondratiev cycle). Certainly a solar economy will entail some adverse environmental impacts, but compared to any of the alternatives, they are minimal indeed. Despite eco-radical fears that PV collectors would monopolize the earth's desert surfaces, careful calculations show that all of this country's electricity needs could be met be devoting only .37 percent of its territory to PV arrays (Weinberg and Williams 1990:149). This is one sacrifice that the earth can certainly afford. As Oppenheimer and Boyle argue, economic and ecological health are mutually supportive, not mutually contradictory. But so long as American environmental protagonists and antagonists continue to regard the two as incompatible, the United States will remain a sorry laggard in the global transition to an ecologically sustainable economic order.

Nanotechnologies
Although Oppenheimer and Boyle present an exciting vision of the environmental possibilities offered by select high technologies, K. Eric Drexler (1986; Drexler and Peterson 1991) offers a far more daring and (guardedly) optimistic scenario of a future society enjoying the fruits of "green wealth." Drexler powerfully argues that molecular nanotechnologies should make virtually all present-day technological forms obsolete, perhaps within the next few generations. "The industrial system won't be fixed," he informs us, "it will be junked and recycled" (Drexler and Peterson 1991:22). In his vision molecular assemblers guided by minuscule nanocomputers will be able to construct atomically precise yet surprisingly inexpensive goods of tremendous variety. A veritable cornucopia of smart materials, able to repair themselves and rearrange their shapes to fit the needs of their users, supposedly awaits just the other side of the impending nanotechnology revolution.

For the Promethean environmentalist, the appeal of nanotechnology lies more in its environmental promises than in its potential to provision human needs and wants. Not only will molecular processing release no pollutants, but molecular devices could be employed for cleansing the earth of its twentieth-century contaminants. Indeed, these very pollutants, especially waste carbon dioxide, should provide nearly the entire resource stock necessary for the new economy. Forestry, fiber growing, and even mining will therefore become obsolete. Drexler even gives hope to the ultimate eco-Promethean fantasy: species restoration. Combining nano- and genetic technologies, he believes, may allow us to recreate extinct forms of life, so long as their genetic codes are preserved in tissue samples. Here one can appreciate how the Procheans' perspective exceeds that of the Arcadians in its ultimate vision of environmental restoration.

Despite its careful grounding in physics, chemistry, and mechanical engineering, nanotechnology is still a somewhat distant dream, and the advances sketched above may never be realized. And even if the visionaries are proved correct, great dangers still await. As Drexler unhesitatingly reveals, nanotechnology could prove a potent carrier of military destruction (see also Milbrath 1989). A certain degree of social control is thus vital, just as it is for other forms of advanced technology. Moreover, nanotechnologies will never allow a complete decoupling of human beings from the natural world, most importantly because they will never yield foodstuffs (molecular devices will not mimic biological structures). As the following discussion reveals, agriculture continues to present some of the most intractable environmental problems.

Agriculture
The environmental dilemmas of agriculture seem especially vexing. The human population has no option but to feed on other living organisms, thereby of necessity monopolizing a large percentage of the planet's primary productivity. Because agriculture necessarily entails the manipulation of ecosystems, decoupling processes are not easily applied. The spatial organization of agriculture also makes pollution control remarkably difficult. Whereas factories spew out waste from a limited number of stacks or pipes, farmers disseminate fertilizers and biocides over a wide expanse of territory. Sophisticated pollution control devices cannot easily be installed where waste seeps from such nonpoint sources.

The eco-radical answer to the agricultural impasse is a return to organic farming. Chemical-free cultivation does indeed have much to recommend it, although if it is to become economically competitive, concerted (and highly specialized) research will be necessary in such areas as integrated pest management (IMP). In the absence of significant IPM advance, increasing production costs will translate into either significantly increased food bills or lowered dietary standards, a situation few Americans would tolerate. In the near term, methods derived from organic farming might be combined with selected new technologies, allowing farmers to reduce their reliance on chemical inputs, especially those that present the greatest environmental hazards. In the Third World especially, such intermediate tech approaches to agricultural production are desperately needed (see *The Economist,* "The Green Counter Revolution," April 20, 1991, pp. 85–86).

Many green extremists, however, deny that anything new is needed. Instead, they point to the agricultural success of the old order Amish, a people who rely on traditional farming techniques, shunning agricultural chemicals and modern machinery (Berry 1977: 210 ff.). What they fail to mention, however, is the fact that Amish patriarchs owe much of their success to their exploitation of the labor of their *numerous* children. If all of our farmers were to adopt an Amish way of life, rural America would begin to resemble rural Bangladesh, both in terms of population density and in regard to patriarchal tyranny, within the span of a few generations.

Yet agro-environmentalist tracts, even those of a radical bent, do contain many worthwhile suggestions. As most argue, the need to adopt a less carnivorous diet is paramount. Meat production is energetically inefficient and ecologically unsound; when cattle convert grain into meat, most of the original food value is lost in metabolic processes. By relying substantially on grain, pulses, and farm-raised fish,[2] we could return vast expanses of agricultural land to nature, reduce our increasingly suffocating medical expenditures, and at the same time drastically curtail our use of pesticides and fertilizers. Eco-radicals are also correct in arguing that small-scale cultivation must persist at some level, if only to preserve the genetic diversity of crop plants. Modern farming *relies* on the diverse array of genetic materials maintained by indigenous farmers, particularly those living in remote Third World villages, yet consistently *undermines* that diversity by disseminating "improved" cultivars. Gardeners in the industrialized nations can do their part by assiduously cultivating "heirloom" fruits and vegetables, and by carefully selecting and exchanging their seeds (Pollan 1991, chapter 11). In agriculture, high-tech approaches are often helpful, but they will never prove adequate.

More innovative ideas from the eco-radical community could also help us devise less destructive forms of agriculture. The geneticist Wes Jackson, for example, daringly argues that we should abandon annual crops, such as wheat, and instead rely on perennial plants that produce year after year (Jackson and Bender 1984). The cultivation of annuals demands plowing, leading inevitably to soil erosion. Although no-till farming practices are now being explored by conventional agricultural researchers, these techniques generally require massive applications of herbicides and fungicides. Jackson, therefore, advocates cultivating perennial grain crops that would require neither constant plowing nor chemical control. The only hitch is that such crops do not yet exist; Jackson and his colleagues are presently working to create them through traditional breeding techniques. A similar and more immediately practical idea was forwarded several decades ago by geographer J. Russell Smith (1953), who urged farmers to reorient their agriculture toward perennial tree crops, such as chestnuts, primarily in order to save the country's remaining topsoil.

Yet while organic farming, reduced meat consumption, and permaculture offer some hope for solving the agricultural crisis, their impact to date has been marginal at best. Organic crops are generally too expensive, and often too imperfect, to appeal to a broad market. Despite a modest reduction in red-meat consumption (due primarily to health concerns), the deep attachment to animal flesh seems too strong to be overcome through moral persuasion. Finally, the perennial grains developed thus far yield insubstantial harvests, while arboriculture remains untenable for both economic and gastronomic reasons.

But these same environmental dreams could perhaps be realized if we were willing to harness technology to the task. Genetic engineering is particularly promising in this regard (Gasser and Fraley 1989).[3] The traditional breeding techniques of artificial selection ultimately depend on the random appearances of desirable genetic mutations; at best they require dozens of plant generations to come to seed before modest improvements can be realized. High-yield perennial grains may someday appear, but probably not until many decades have passed—a time span we cannot afford. Through recombinant DNA, on the other hand, "designer" organisms can often be created in months. The careful application of biotechnology to other agricultural problems offers further environmental advantages. Organic farming, for example, will receive a tremendous boost as geneticists fabricate crops that manufacture their own internal pesticides. Similarly, fertilizer inputs can be drastically curtailed once genes for nitrogen fixation can be inserted into non-leguminous crops plants.

As advances in biotechnology make agriculture more efficient, large tracts of land can be progressively returned to nature. Similarly, intensive greenhouse cultivation, relying on high-tech glass construction, advanced atmospheric chemical control, and perhaps even the use of molecular antifungal agents, could increase food supplies while at the same time tremendously diminishing the extent of land needed for food production (Drexler and Peterson 1991:175). Yet some American politicians appear to rule out such possibilities beforehand, assuming that increasing production will only translate into larger commodity gluts (Sagoff 1991:353). Certainly the biotechnology revolution will require a difficult set of adjustments for American farmers, but only an anti-environmentalist would automatically rule out the possibility of reducing the extent of land monopolized by agriculture. Agricultural gluts represent political, not technological, failure.

Advanced techniques in food science, especially those concerned with enzyme production and protein synthesis, may also offer substantial environmental benefits. Especially desirable is the development of palatable, vegetable-based meat substitutes. If soy burgers become indistinguishable from, and less expensive than, the genuine product, we could expect widespread cutbacks in meat consumption, allowing us to liberate vast tracts of land from agricultural production. Such environmental benefits would, however, be impossible to realize if consumers were to take at face value the eco-radical tenet that artificial products are to be avoided in all instances.

Radical environmentalists will likely respond to the proposals sketched above with disgust if not revulsion. In their view, tampering with DNA is blasphemy, and even the consumption of artificial foods is something of a venal sin. But by sanctifying the human place within the natural world, radical greens only ensure the destruction of nature. The more we feel compelled to consume natural products, the more we monopolize the earth for ourselves.

The eco-radical denunciation of genetic engineering also betrays a misunderstanding of our historical relationship with the natural world. We commenced playing God millennia ago, as soon as neolithic humans began to domesticate plants and animals. There has never been, for example, a single stalk of wild corn; maize was not domesticated so much as created by the crossing of different wild plants that would never have shared their genes without human meddling (Heiser 1981:107). The primitivists, who do grasp this truth, conclude that agriculture represents our original sin. Perhaps it does. Yet I continue to believe that we can best atone for our past environmental crimes not by retreating toward an unreachable Arcadian past, but rather by moving forward into a benign Promethean future.

Of course, genetic engineering, like other forms of high technology, can certainly be misapplied. One current project that borders on insanity involves the development of a herbicide-resistant strain of tobacco (Gasser and Fraley 1989). This will only offer the world a more abundant supply of an addictive, deadly drug—as well as a more poison-filled environment. Genetic technology, like all others, requires firm political and moral guidance.

The proposals sketched above may offer hope for the long term, but for the short term more immediate steps must be taken. American agriculture is indeed in a crisis situation, which has very dangerous environmental implications. Heavily indebted farmers are forced to expand recklessly in order to ensure harvests large enough to cover their interest payments, a situation that leaves them no room in which to experiment with ecologically sound alternative methods. Because of its intimate connections with nature, farming cannot be considered just another economic activity, and the market certainly cannot be relied upon to generate solutions to the current impasse. Unfortunately for the consumer, somewhat higher prices for agricultural commodities are probably necessary if American farmers are to receive the breathing room they so desperately need. We must begin to break our addiction to chemical farming—a process that will entail some pain for society at large.

Conclusion
The development of ecologically forgiving technologies is not inevitable. Desirable advances can only be realized through great efforts undertaken by large segments of human society. Americans should devote unyielding efforts to enhance education, scientific research, and economic productivity. If present trends continue, any fifth wave of economic growth will be dominated by Japan, not the United States. It would not bode well for either human freedom or environmental protection if the United States were simply to abandon the effort. Yet the chances of American leadership in the development of an ecologically sustainable socioeconomic order seem slim indeed. Both eco-radicals, who despise capitalism and denigrate technology, and anti-environmentalists, who worship at the alter of the free market oblivious to environmental destruction, seem perfectly willing to watch the United States shed all its competitive advantages. As Porter (1990:173) shows, nations either move ahead or fall behind in international economic competition. And as Mokyr (1990) demonstrates, the historical reality is that the forces of conservatism—in this case, including both the extreme right and the eco-radical left—more often than not thwart the development of promising new technologies, even in societies that were once technological leaders.

Technological advance has clearly been something of a two-edged sword. The vast majority of people in preindustrial times may have lived short and impoverished lives, but industrialization has brought us face to face with global warming, ozone depletion, and acid rain. Given this trade-off, most green radicals would conclude that ecological salvation is more important than human comfort or longevity.

There are two fundamental problems with this line of reasoning. For one thing, it fails to recognize that industrial pollution is only one kind of environmental degradation. Preindustrial peoples have proved themselves capable of extraordinarily destructive acts, notably by deforesting entire landscapes and exterminating major faunal species. More importantly, the antitechnology thesis ignores the fact that technological advance has the power to heal as well as to destroy. In the modern world technological poverty often forces immiserated peoples to degrade their environments. Similarly, old industrial processes are virtually synonymous with dirty industrial processes. I am convinced that we can develop a clean, environmentally benign industrial system, but only if we have the will to embrace technological innovation and support the educational infrastructure that makes it possible. And despite the claims of *all* eco-radicals, such a transition will only be possible if we retain a capitalist economic system.

5

The Capitalist

Imperative

■ The Radical Position

The Impact of Marxism

The belief that capitalism is a root cause of our environmental crisis is common to all varieties of radical environmentalism. Among eco-marxists capitalism is, of course, singled out as the overwhelming source of all ills, social and environmental. But even nonmarxist greens like Herman Daly insist that an ecologically sustainable, fully capitalist economy could not exist and that an alternative economic system must therefore be devised (Daly and Cobb 1989:2).

While marxian ideas have been all but ignored in the three previous chapters, here they occupy center stage. Eco-marxists, in fact, share few of the concepts criticized in the previous chapters, and they are in general the most theoretically sophisticated of radical environmentalists. Marxism's criticisms of capitalism are, moreover, far more detailed and voluminous than those of competing eco-radical philosophies. Accordingly, the objections to capitalism found within even the anti-marxian varieties of radical environmentalism are strongly influenced, directly or indirectly, by marxist thought.

Because marxism's influence on the popular imagination in the United States (unlike in much of Western Europe) has remained marginal and indirect, this chapter is concerned largely with academics who write for academic audiences. Considering the philosophy's recent global retreat from political power, one might conclude that the following pages merely flog a dead horse. Yet academic marxism is far from dying. In one form or another, the doctrine has successfully appropriated the intellectual and moral high ground in a variety of fields, and it shows little sign of ebbing, despite recent events on the political front. Although economics and political science have been little affected, sociology, anthropology, literature, history, and geography have all seen the development of powerful marxian contingents.[1] Similar developments are also apparent in regional studies, where a sizable majority of Latin Americanists and Africanists find marxian theories compelling if not completely convincing.

Despite its academic stature and its pervasive impact on all species of radical thought, marxism as an explicit political philosophy finds little favor in most eco-radical communities. If the spectacular ecological and social failures visible in all communist and ex-communist countries is not dissuasion enough, many environmentalists are ready to dismiss marxism as irredeemably sullied by its humanistic heritage. Yet while denouncing marxian doctrines, radical greens retain, and indeed often extend, the marxian critique of capitalism. Some carry this critique to all existing political regimes, including so-called communist ones. According to many eco-radicals, the Soviet Union represented an ultimate form of monopoly capitalism in which the sole surviving corporation had fully merged with the state (Bookchin 1989:128).

Equating capitalism with contemporary communism might seem to require remarkable intellectual legerdemain, but even a few self-identified marxists have found it a compelling gambit; Deleage (1989: 25), for example, dismisses the USSR of the 1980s on the grounds that it was guided by the "imperatives of state capital" (see also O'Riordan 1989:78). More often, academic marxists simply dismiss the modern socialist and exsocialist industrial nations as bureaucratic deformities that lost the true marxian vision, allowing them to disown the social and ecological catastrophes of the former Soviet Union, Eastern Europe, and China. Yet this remains a tricky issue, as few wish to relinquish entirely the great communist experiment. Many writers both tentatively defend and simultaneously criticize (recently) existing socialist states (for example, Johnston 1989). In the end, however, the environmental and social failings of a country like the Soviet Union prove to be rather beside the point. Existing capitalism, not hoped-for communism, is the primary subject of virtually all academic marxist works. While the modern green-reds promise to include socialist ecological failures within their analytic purview, "capitalism's global destruction of nature" remains their over-riding concern (J. O'Connor 1989b:10).

The essential aim of eco-marxism is to use the environmental crisis to revitalize marxism. In practice, this involves two procedures: first, showing that whereas communism may be *accidentally* destructive, capitalism *necessarily* destroys the earth; and second, rehabilitating Marx and Engels as ecological thinkers. The initial task is easily accomplished. All environmental horrors in the First and Third Worlds can, in one way or another, be linked to capitalism, and imaginative thinkers are even able to blame ecological disasters in the former Soviet Union on the capitalist world system, if not Russian-dominated state capitalism. The latter task, however, is not so simple.

In their voluminous writings, Marx and Engels did embrace certain conservation principles (J. O'Connor 1989b:9), but in a manner that most modern greens would regard as hopelessly anthropocentric. There is little if any room for the struggle to preserve wilderness for the sake of wilderness in the marxian agenda. Most varieties of marxism have not only downplayed nature but have largely ignored rural issues as a whole, at least in the industrialized world (Fitzsimmons 1989:113). Nonetheless, the flexibility of contemporary marxism is such that green-red thinkers can still set their sights on capturing the leadership of the entire radical environmental community. Indeed, James O'Connor (1989b:13) describes the movement as "a 'fifth international' as it were."

The following discussion outlines the basic charges against capitalism expounded, often implicitly, within all versions of eco-radicalism. The resulting picture is of necessity greatly simplified; in places it verges on caricature. To balance the simplifications needed to create such a composite view, a subsequent section explores some of the main points of disagreement and debate within the contemporary academic far left. While these debates stray at times quite far from ecological concerns, it is essential to reconstruct in some detail the intellectual underpinnings of the critique of capitalism that in one way or another inform all varieties of eco-radical thought.

The Case against Capital
The most immediate denunciations of capitalism focus on its profit-seeking orientation. Capitalist firms are seen as avaricious in their pursuit of gain, ruthlessly exploiting workers, consumers, and the land so that owners can accumulate ever growing piles of wealth. Capitalism is depicted as an immoral system, based solely on greedy self-interest. The minions of capital are precluded from possessing any sense of ethics or responsibility, as this would only hinder their mandated task of enriching the shareholders. While the logic of capital alone is thought to demonstrate the system's exploitative nature, radicals can easily mar-

shal empirical evidence demonstrating countless examples of corporate brutality.

As discussed in chapter three, eco-radicals also believe that capitalism leads inexorably to ever greater centralization, hence to ever more exploitative social hierarchies. As firms compete, the most avariciously competent will destroy all others. Power thus concentrates in ever fewer hands, destroying in the process any hope of realizing genuine human dignity and freedom.

According to classical marxist theory, capitalism is structured around an invidious twofold distinction in human society, one separating those who own the means of production from those who control nothing but their own labor power. The former group is wealthy and domineering, yet does no productive work; the latter, who actually create all value through their toil, are dominated, impoverished, and oppressed. Because labor is the sole wellspring of value, investors are by definition parasitic. No matter how much workers are paid, they are still taken advantage of, since the owners of capital siphon off the surplus value that only labor can create. Capitalism is also said to demand the maintenance of a "reserve army of the unemployed" that firms can tap when the business cycle swings into full gear. Massive unemployment is thus endemic to the system, bringing untold hardship to the working class. Finally, modern radicals see capitalist exploitation as marching hand in hand with racist and sexist social attitudes. White men, who run the capitalist machine, find great advantage in oppressing women and people of color, as this provides them with a large force of cheap and pliant labor.

The binary structure of capitalist domination is seen as replicated at a global scale. According to the neo-marxist theory endorsed by most eco-radicals, the world is divided into two camps, a wealthy capitalist core, and an impoverished, resource- and labor-providing periphery. Although orthodox marxists diverge here, many contemporary radicals view the success of the West as the singular result of its rapacious exploitation of the Third World, rather than of any internal economic dynamic, let alone virtue (see also chapter six).

Because of the vast power of the capitalist owners and their firms, most eco-radicals consider bourgeois democracy a sham. Corporations are believed to brainwash the populace into accepting the base ideology of individualism through their lock hold on the media. The powerful thereby exercise a cultural hegemony that allows them to manipulate alienated workers into conspiring in their own exploitation. "Commodity fetishism" becomes rampant under capitalist relations of production, as workers, bought off with the trinkets of mass production, are blinded to their own condition. Meanwhile, elected bodies become little more than stooge-councils for the monied powers that have learned to orchestrate the public through lies, manipulations, and behind-the-scenes power plays. Genuine or popular democracy is considered incompatible with capitalism; according to one highly respected marxist geographer, true democracy "can only take place in a classless society produced via the dictatorship of the proletariat" (Johnston 1989:201).

Underlying capitalism is the market system. While a given society may employ market mechanisms to allocate certain goods and services without being fully capitalistic, capitalism itself is inseparable from market organization. The market is alleged to be an inescapably oppressive means of distributing resources, as it allows the wealthy to purchase whatever they please while the poor remain free to do without. Moreover, markets are inherently unstable, wreaking havoc on the vulnerable working class. When the prices of necessities rise more rapidly than do wages, the poor find themselves squeezed in a relentless vice of economic privation.

Most contemporary radicals deny that markets are a natural outgrowth of the human propensity to truck and barter. Instead, they view them as political creations of relatively recent vintage (Walker 1988). Precapitalist economies, it is often averred, were based either on culturally embedded personal gift transactions or on state-instituted set-figure exchanges, but never on impersonal market haggling (see Polanyi 1957). Since the market is a recent sociopolitical invention, there is no reason why it cannot be eliminated and replaced by a more just and stable mode of allocation. Moderate environmentalists' proposals to harness market mechanisms for pollution control are thus greeted with derision by almost all radical greens.

The culmination of the left-radical critique is the conclusion that capitalism is so beset with contradictions that it will eventually self-destruct. Classical marxism locates the main blockages in the machinery of accumulation, especially in the supposed propensity of the profit rate to fall continually. Contemporary green radicals, on the other hand, are more concerned with environmental contradictions, especially the seeming impossibility of continued economic expansion on a finite planet. By incessantly fouling its own nest, capitalism will surely undermine itself. Indeed, some radicals claim that the costs of environmental regulation helped spark the economic crises of the 1970s and 1980s (Faber and O'Connor 1989).

Finally, both traditional marxists and modern eco-radicals share an eschatological assessment of the world, although they often see the period following the final days of capitalism in strikingly different terms. Whereas the classical marxist vision foretold the relatively swift appearance of a millennial worker's paradise, most environmental radicals fear that as capitalism collapses it may take the entire planet with it. Eco-marxists may thus find opportunity in harnessing the ultimate optimism of the communist vision to a movement long hobbled by its profound pessimism.

But if eco-marxists retain their hope for an eventual communist utopia, in the short term they see only evidence of advanced decay everywhere they look. Crisis is ubiquitous in the United States; our environment is perishing, our economy is falling apart, and our society is becoming more derelict year by year. Any signs of economic stability, let alone growth, are chimerical. As O'Riordan (1989:77) informs us, by the 1980s "most U.S. wealth was borrowed." Even the modern personality is in crisis; according to James O'Connor (1987:179), the leading proponent of eco-marxism, "The lid is ripped off the id. Libidinal energy becomes commodified, reified in pop music videos." Throughout the world, the enlightened observer can witness capitalism consuming and destroying the very social fabric and natural environment that make its existence possible in the first place. The end of "late capitalism" may well be in sight.

Marxists have, of course, been predicting the imminent demise of capitalism for over 100 years. Contemporary believers, therefore, must explain how the bourgeoisie have been able to devise temporary expedients (sometimes termed "fixes") that allow the moribund system to survive a while longer (see Harvey 1982). Thus, through "spatial fixes" capitalists redeploy their operations to the far reaches of the globe, finding new terrain for profit making once the inherent contradictions of capital have exhausted the possibilities in the old. The amount of space available for the capitalist machine to arrogate, however, remains finite. While contemporary marxists are increasingly inclined to admit that capitalism shows vast adaptive powers, few spurn their doctrine's basic postulate that it must eventually fail.

Ecological Contradictions

Ecological contradictions within capitalism have most often been emphasized by nonmarxists, but this is precisely the grounds on which eco-marxists now seek to recuperate the communist endeavor (M. O'Connor 1989; Deleage 1989). An initial problem they isolate is the fact that the anonymity of the marked hides the environmentally destructive consequences of commodity production (Peet 1991: 516). Capitalism's *ines-*

capable failing, however, is that it glorifies—and indeed, requires—continual economic growth; only when expanding can a capitalist economy be considered healthy. Such expansiveness is cancerous, entailing the inexorable consumption of finite natural resources. Eventually the system will encounter the earth's own limits, at which point it cannot help but collapse. The final reckoning, however, may be delayed by technological innovations. Substitutes, for example, can be discovered for many diminishing resources, providing threatened industries with another breathing spell. But, as even the nonmarxian and relatively moderate writer Paul Ehrlich soberly informs us, "the real opportunities for adequate substitution are limited" (Ehrlich 1989:13).

According to the noted marxist geographer David Harvey, capitalism's ultimate environmental contradiction stems from its short-term time horizon. Neoclassical economic thought is founded on the idea that future benefits are of lesser value than benefits that can be immediately realized; a discount rate (related to the interest rate) is thereby used to downplay the future advantages that might be gained by conserving resources. Environmentalism, in contrast, is founded on a long-term, multigenerational (if not geological) time horizon, one in which the need to preserve resources becomes paramount. Harvey concludes that what distinguishes the environmental movement "is precisely the conception of time and space which it brings to bear on questions of social reproduction and organization" (Harvey 1990:421).

In the early 1970s technically oriented environmentalists devised computer simulations to prove that the global economy would soon collapse as essential resources, such as copper, neared exhaustion (Meadows et al. 1972). At the time, several marxist scholars attacked such methods, as well as their underlying neo-malthusian implications, fearing that they were but a mask craftily donned by a metropolitan elite who wanted only to continue monopolizing the earth's resources. In the current reassessment offered by eco-marxism, such "scientistic" methodologies are still regarded with grave suspicion, but a cautionary stance toward resource depletion has finally been adopted. What was considered highly reactionary only fifteen years ago has somehow been transformed into something that can now be regarded as very radical indeed.

A sustainable economy, eco-marxists tell us, can only be realized by instituting socialism. What they seem to envision is a state along the lines of contemporary Cuba, although presumably without the personality cult, a little less bureaucratic in orientation, and lacking the ideology of productivism. The nonmarxian majority of eco-radicals, as we have already seen, also argue for socialism of a sort, but they remain convinced that socialism can never be anything but a mirage unless we return to small-scale, anarchistic communities. The more moderate, compromising fringe of the eco-radical movement hopes to retain a large-scale market system of some sort, but agrees that capitalism as we know it is simply too expansionary and too exploitative to persist (Daly and Cobb 1989).

The basic marxian critique of capitalism, as sketched above, exerts a powerful influence over all schools of eco-radicalism. But beyond this there is little common ground. Even within the compounds of explicitly marxian scholarship, disagreements are frequent and occasionally fierce.

The Varieties of Marxian Experience

The global collapse of communism is an obvious threat to the marxian doctrines upon which the eco-radical critique of capitalism rests, yet many marxist scholars would hope to turn this crisis to their own advantage. Eastern European socialism has long been a major embarrassment anyway, and many would just as soon relinquish the tasking job of defending it.

Not surprisingly, different marxist schools offer varying responses to the current challenge, just as they have long offered disparate explanations of communism's decidedly nonparadisiacal incarnations. A common, although hardly materialist, approach has been to dismiss Eastern Europe's errant path as the result of historical contingencies and personal whims, Stalin's accession to power usually being held as the pivotal accident. Others have argued that the machinations of the capitalist powers have forced the poor, struggling socialist states to devote too many of their resources to the military, thus forestalling the development of the economic base necessary to begin constructing real communism. A more encompassing thesis is that in a *world system* dominated by capitalism, true socialist reform becomes difficult to undertake, simply because would-be socialist states do not have adequate room in which to maneuver (Peter Taylor 1989:347). Other arguments are more simplistic: Commoner (1990:220–21), for example, seems to blame the Soviet Union's ecological disaster largely on the fact that it imported its basic industrial techniques from the West.

A few eco-marxists, however, are beginning to question their own visions of a classless future, asking how an economically successful, humane, and environmentally benign form of socialism might be devised. But this has seldom advanced beyond the stage of idle wondering; few marxists yet have much to say about communism. They aim their analyses squarely at capitalism, and they define their own position as that of critics of the current order rather than visionaries, let alone planners, of a new society. One searches in vain for concrete proposals for constructing a democratic, prosperous, and ecologically sustainable form of marxian socialism.

One reason for marxism's failure to develop a coherent theory of socialism is its underlying proposition that virtually all problems stem from capitalism alone. Capitalism is pictured as a totality, the one hideous reality that structures everything in its own image. According to James O'Connor (1987:156), only "tiny corners of social life [are] not yet colonized by capital and the state administration." Thus, once capitalism is *globally* extirpated, many believe, a healing process will inevitably commence. But that is a task for later days; the immediate job is to analyze, and then seek to undermine, bourgeois civilization.

Academic marxism may center around the analysis of capitalism, but it is a grave mistake, typical of those who denounce marxism without first intellectually grappling with it, to suppose that any single theory of capitalist dynamics prevails. Orthodox marxism of various stripes, the several neo-marxisms, neo-orthodox marxism, structuralist marxism, post-structuralist marxism, post-enlightenment marxism, postmodern marxism, analytical marxism, and the French regulation school (let alone leninism, trotskyism, and maoism), offer divergent explanations of the capitalist system. The more innovative recent versions are highly sophisticated, having abandoned the strained notions that bourgeois society consists of two classes only (owners and workers) locked in internal conflict, and that capitalism is destined by its internal logic to perish, yielding in the process to a virtuous proletarian regime. Although traditional marxism is dogmatic, deterministic, and teleological, one cannot level such accusations against some of the new, refurbished varieties (Corbridge 1988). Moreover, these various marxian schools can be highly contentious, struggling to claim for themselves, and deny the others, the mantle of Marx and Engels.

In devising a more sophisticated socioeconomic theory, contemporary heterodox marxists have strayed quite far from Marx's original vision, leading several outside observers to wonder just how much marxism one can dispense with and still label one's self a marxist (Bell 1990:310). Indeed, nonmarxists' social theories, such as those propounded by Anthony Giddens, are often "closer than many contemporary Marxists' to orthodox Marxism" (Wright 1983:11). But in becoming more philosophically open, nonorthodox marxists do not necessarily become any less radical. To the contrary, many contemporary neomarxists fiercely deny that capitalism can ever be considered progressive (for example, Peter

Taylor 1989:346), thus jettisoning Marx's fundamental assertion that socialism must be presaged by the development of the forces of production that can occur only under a regime of capitalist accumulation.

Nonmarxist eco-radicals, it turns out, almost universally agree with this hyper-marxist assertion that capitalism never has any positive qualities. They also not uncommonly misconstrue capitalism in a manner that might make a careful marxist scholar blush. Roderick Nash (1989: 201), for example, feels that the severity of American slavery can be explained as an instance of unmitigated capitalism. Although marxists often argue that a slave "mode of production" can "articulate" with, and thus benefit, a capitalist one, they certainly recognize that free labor is a *defining* characteristic of capitalism as a system.

Because of this conceptual inconvenience, the radical environmental historian Donald Worster now argues that the conventional definitions of capitalism are too narrow (1990:1,098). It is, after all, easier to denounce capitalism if one can include all objectionable social forms under its umbrella. For this advantage, a loss of theoretical precision may seem a reasonable price. If one is willing to ignore facts, even greater gains can be achieved; Worster (1990:1,106), for example, informs us that the rainforests of Borneo have passed "to modern corporate ownership," evidently unaware of the fact that the ownership of these forests resides, socialistically, with the states of Indonesia and Malaysia (see Repetto and Gillis 1988) (for a telling critique of Worster's position, see Cronon 1990).

Heresy within the Ranks

Eco-radicals thus concur that capitalism is highly destructive of both nature and society, but they disagree strongly when debating its internal dynamics, its historical role in socioeconomic evolution, and its future trajectory. Assailing the radical environmental theory of capitalism, (whether explicit or implicit), thus becomes a futile exercise, since one will immediately be counterattacked for taking on a straw-person or for naively grappling with an outdated and entirely passe (if not simply vulgar) species of anticapitalist thought. Critics of marxism are also commonly dismissed as mere bourgeois thinkers, the implication being that they are necessarily mystified by capitalism because of their own self-interest in the system's perpetuation.

If marxists can casually dismiss most competing theories as self-interested and collaborationist, they face a special challenge when confronting defectors from their own ranks, some of whom have recently defined themselves as post-marxist (see Corbridge 1986, 1988, 1989,

1990). Because of their thorough schooling in marxian debates, post-marxists are much better prepared to grapple with marxism on its own multifaceted terrain than are garden-variety bourgeois scholars. Their general thrust has been to accept the political-economy framework of marxian studies, but to reject its certainties and its exclusion of competing theories. They are characteristically cagey about their own ultimate political beliefs; as the brilliant geographer Stuart Corbridge writes: "[Post-marxism] is not opposed to socialism . . . but it is concerned to theorize each enterprise and to concern itself with the contours and contradictions of actually existing socialism" (1989:245). Corbridge does imply, however, that genuine (and redeeming?) reform is possible within capitalism, and he insinuates that progressives should strive to work within the existing system (1990:634). In a similar vein, Piers Blaikie reminds fellow socialists that capitalist enterprises can be socially progressive, that the presence of transnational corporations is often a necessity in poor countries, and that environmental degradation can be remarkably severe in marxian states (1989:135, 142, 133). And finally, Andrew Sayer has warned the marxian academic community that it must take seriously the fall of communism in Eastern Europe, since that system was undermined by problems of coordination and motivation for which "Marx himself bears no little responsibility" (Folke and Sayer 1991: 242).

The marxian response to post-marxism has varied from puerile outrage (Blaut 1989), to cajoling attempts to bring the errant children back into the fold (Watts 1988), to open-armed embrace. The last alternative, however, may bespeak a certain naivete regarding the post-marxian agenda. James O'Connor (1989b:4), for example, seems to regard post-marxism as a user friendly version of the essential doctrine; I see it rather as the potential undertaker of the entire effort. Warier marxist scholars, on the other hand, greatly fear the "growing chorus of criticism of marxist scholarship within the left" (Walker 1989:133). One prominent response is a desire to circle the wagons; geographer Richard Walker (1989), for example, calls for peacemaking within the broadly marxian left, coupled with the concerted exclusion of all scholars who deny the fundamental marxian core. Indeed, he goes so far as to berate the editor of a "left" scholarly journal for having the temerity to publish a liberal critique of marxist scholarship (Walker 1989:160).

Walker is certainly correct in identifying threats from the left as the main challenge to academic marxism. Liberal (let alone moderate or conservative) critics can easily be ignored, but what of the radical feminist who illustrates the patriarchal underpinnings of contemporary marxist thought? The adherents of a vague school that might be labeled "subversive postmodernism," a group drawing inspiration from the deconstruction movement in literary theory, have been known to attack fiercely the monumental certainties of mainstream marxism. To a radical postmodernist feminist like Gillian Rose (1991:120), orthodox marxism (as exemplified by David Harvey [1989]) "embod[ies] the characteristics of western masculinity: hard, logical, certain, oppressive." Harvey (1989:350), for his part, has accused the deconstruction movement on which much of postmodernism rests of having "produced a condition of nihilism that prepared the ground for the re-emergence of a charismatic politics [that is, fascism] and even more simplistic propositions than those which were deconstructed."

The feminist criticism of marxism has much to recommend it; there are indeed close connections between all forms of totalitarianism and traditionally masculine thought. But at the same time, by embracing a self-consciously subversive form of politics that fully endorses the marxian denunciation of capitalism, writers like Rose marginalize their own position. Whereas post-marxism tries to redirect academia's leftward impulse back toward the center, subversive postmodernism strives to deflect it wildly in ever more radical directions. Here, I would argue, lies great danger; irrationalism may be inherently radical, but it can just as easily be harnessed to the radical right, as the examples of the philosopher Heidegger and of the deconstructionist savant Paul de Man—onetime nazis both—so clearly show (see Lehman 1991).

But more often, one must admit, philosophical radicals become, in practice, political ciphers. As David Lehman (1991:70) cogently writes: "Deconstruction makes possible, moreover, a risk-free form of subversiveness. It gives its adepts a way to look daring while playing it safe—to mouth the rhetoric of the radical while climbing up the tenure ladder to pluck the fruits of the system whose legitimacy they claim to question." Much the same could be said about many academic marxists.

Grounds for Comparison
The remainder of this chapter largely bypasses the internal debates within the radical left on the nature of capitalism; for such an account one should turn to the writings of Corbridge and other post-marxists. I do not feel compelled to address these issues in part because to do so would be to engage the debate on marxian terrain. Here the non-marxist is on perilous ground indeed, since the arguments have been structured beforehand to ensure the defeat of capitalism. In a tactically brilliant but intellectually indefensible gambit, marxists have insisted that cap-

italism be judged by its most egregious practices, whereas marxism is to be evaluated according to its diverse and never stationary critique of capitalism.

In contrast, the stance taken here is that both capitalism and marxism must be assessed by the same criteria. In particular, we should examine how each system has performed in practice, and we should explore the potentialities of each system for achieving environmental sustainability and social justice. On the former score, capitalism—for all its faults—is clearly preferable. In regard to the latter issue, marxism begins with an initial advantage deriving from its utopian visions. But until marxist thinkers begin to devise blueprints of how "true" socialism might be achieved, one is forced to regard those visions as jejune fantasies. Capitalism, on the other hand, has historically demonstrated vast potential for real social and environmental reform, while potentially workable designs for further amelioration have been forwarded by numerous liberal scholars.

Although it is essential to realize that there are many different variants of capitalism (some of which are much more socially and environmentally responsible than others), the unbridgeable gap separating capitalism from marxian socialism must also be recognized. A few leftists may now be touting market socialism, while many environmentalists have long wished for a convergence between capitalism and socialism (for example, Scheffer 1991:170), but such hopes rest on a profound misunderstanding, one that orthodox marxists, at any rate, would never make. A government that mandates social security, for example, does not thereby become partially socialistic, at least in the marxian sense of the term. As Peter Berger cogently insists (1991:xiii), capitalism and socialism must be seen as systems of production, not distribution; Sweden, by this definition, is an overwhelmingly capitalist country that merely has a large welfare system. Unfortunately, many of capitalism's most fervent apologists are equally confused on this score. Arthur Seldon (1990:10, 23), for example, seems to think that Sweden is half capitalist and half socialist. As a result of such errors, writers like Seldon misidentify capitalism with minimal government, a postulate that can only be accepted if one blinds one's self to capitalism's emergent leader: Japan.[2]

Although the following pages unceasingly assail the marxist project, it should be made clear that I hold the greatest respect for numerous marxist scholars. E. P. Thompson and Eric Hobsbawm in social history, Perry Anderson in comparative sociology, Eric Wolf in anthropology, and David Harvey and Edward Soja in geography have presented brilliant analyses of specific socioeconomic processes within the capitalist world.

Indeed, within my own discipline of geography, many if not most of the best scholars in one way or another identify with marxian goals.

Yet the future of American marxism is cloudy, even if it continues to gain strength within academia. Indeed, the very invention of an explicitly environmental marxism can be read, in part, as an attempt to gain broader support for a floundering movement. Such a tactic is clearly evident in the recent greening of Western European communist parties. But while eco-marxism may in the future thrive in select American university departments, I seriously doubt whether it will have much impact on the larger eco-radical community, a group expressing massive fear of any form of state power. But regardless of eco-marxism's actual success, the marxian critique of capitalism has had a pervasive impact on all eco-radical scholars, even on those who shun the label and who imagine an anarchistic rather than a communistic future.

■ The Failings of the Radical Position

"Real Existing Socialism"
The easiest defense of capitalism is simply to contrast it with existing and recently existing examples of marxian socialism. As is now abundantly clear, marxism's record is dismal on almost every score, be it economic, social, or environmental. These failures cannot be dismissed as errant quirks; marxian regimes have come to power in numerous countries, and everywhere the results have been disheartening. From impoverished African states like Mozambique, Ethiopia, Guinea, Madagascar, and the Congo to highly industrialized, once-prosperous European countries like the former East Germany and Czechoslovakia, all marxist experiments have ended in disaster. Chapter six will address the failings of marxism in the Third World; the present discussion is concerned with the formerly communist industrial states of Eastern Europe. For convenience sake, the analysis focuses on conditions that pertained before the democratic revolutions of the late 1980s and early 1990s.

Radical greens admit that environmental conditions in Eastern Europe are as bad as those found in the West. But such admissions are far from adequate; by almost every measure, the communist environment is more severely degraded than the capitalist environment. Only with the recent downfall of marxian regimes has the ecological debacle of the East come to light. As our knowledge increases, the environmental conditions of Eastern Europe are revealed as ever more horrific. And when one considers the poor performances of the economies that have wreaked such destruction, the comparison between capitalism and communism becomes one-sided indeed.

Although the general state of environmental devastation in Eastern Europe is now well known, a few specific examples are still in order. It is quite possible that the world's most industrially devastated landscape is that of Poland's Silesia, an area in which the soil is so lead-impregnated as to render farm products virtually poisonous. Nor are conditions much better in other Polish regions. Many Polish rivers are so filthy that their waters cannot even be used for industrial purposes. As Fischoff (1991:13) reports, "by U.S. and European standards, the country has virtually no potable water." In Poland's industrial belt, air pollution, especially sulfur dioxide contamination, far exceeds anything found in the West. Many buildings in Cracow are simply melting away in an acid bath.

Devastation of similar magnitude may be found in many regions within the former Soviet Union. Latvia, for example, is burdened by many poorly regulated and constantly oozing toxic waste pits, and its Baltic shores are heavily contaminated with bacteria, heavy metals, and even chunks of phosphorus (in 1988 the Soviet army dropped 400 bombs containing 20 tons of phosphorus into the Baltic Sea [Burgelis n.d.:7]). The transformation of the once-rich Aral Sea into a shrunken, almost lifeless sump is now a virtual international emblem of the powers of human destructiveness (Kotlyakov 1991). Everywhere one looks the stories are the same, recounting one ecological disaster after another.

Equally telling are comparative figures on energy use. One of the principle reasons for Eastern Europe's environmental catastrophe is its appallingly inefficient use of energy. As *The Economist* (February 17, 1990) reports: "On average, the six countries of Eastern Europe . . . use more than twice as much energy per dollar of national income as even the more industrialized countries of Western Europe. Poland, with on some counts a GDP smaller than Belgium's, uses nearly three times as much energy; Hungary, whose GDP is supposedly only a fifth of Spain's, uses more than a third as much energy." Here one can appreciate the environmental consequences of an economy that has approached the vaunted steady-state; lacking economic vitality, the East has been forced to retain an antiquated, inefficient, and highly polluting set of industrial plants. Factories have remained in operation that would have been shuttered decades ago in the West.

The dismal environmental conditions of the communist world stem from the political and economic structures implicit in marxism and not, as academic marxist apologists would have it, from either historical contingencies or the structural power of the capitalist world system. As has been widely noted, under a "dictatorship of the proletariat" (which in practice has proved to be a dictatorship to be sure, but hardly one of the working class), independent activist groups seeking environmental protection enjoy a precarious standing at best. During Eastern Europe's long marxian night, only a few feckless scientific organizations could dare even ask for environmental consideration.

More intractable problems derive from marxist economic philosophy, especially from the belief that labor is the sole source of value. As Rolston (1989:76) writes, "Marxists often argue that natural resources should be unpriced, for in fact resources as such have no economic value." Although marxian regimes never actually distribute natural resources at no cost, they do consistently undervalue oil, timber, and other such materials. By assigning extremely low prices to natural resources, marxist economics ensure that they will be wastefully employed, leading inevitably to needless environmental degradation (see, for example, Barr 1988 on Soviet forestry). Finally, as is widely appreciated, any large-scale economic system that dispenses with the market must instead rely on *command* for fixing the price structure of goods and services. Yet no government command center can ever obtain adequate information to avoid production and distribution bottlenecks. Command economies are thus inherently inefficient, generating economic waste that is invariably linked with environmental degradation.

Other reasons for the dismal environmental failure of communism stem not from marxist ideology so much as from the specific ways in which marxist political parties have stimulated production. Since they deny the profit system, other methods of motivating workers and managers have by necessity eventually been devised. The most common system has been for central planners to set production quotas, and then to reward plant managers who exceed them. As it turns out, the production quota system supplies incentives to ignore existing environmental regulations every bit as powerful as those of the profit system. Other forms of "motivation" have been far more sinister: Soviet geographer Ruben Mnatsakanyan notes that the vast power of the destructive Soviet Ministry of Land Reclamation and Water Management stemmed from the fact that it was originally "a KGB department that dealt with the digging of canals by prisoners" ("The Changing Face of Environmentalism in the Soviet Union" 1990:5).

Eco-marxists often blame the sorry state of real existing socialism on the crucial mistake of bureaucratization. By vesting too much power in the hands of central party functionaries, this line of reasoning goes, the revolutionaries betrayed their own vision of a just, democratic, socialist future. But as Theodore Hamerow's (1990) masterful account of the "graying of the revolution" makes abundantly clear, the rise of the bureaucratic oligarchy may have been unintended but it was nonetheless inexorable. If eco-marxists were ever to gain power in the United States, we could expect history to recapitulate itself on this score.

Still, marxist apologists will continue to inform us that communist leaders just made a few critical errors, and that if we were once again to begin building communism, this time we could get it right. This position might be reasonable had the world known only a single marxist state, but the sad fact is that the experiment has been just as disastrous on every occasion and in every social environment in which it has been attempted. Scholars seeking real material and structural explanations in history would be forced to admit that marxism's political failure has been rather more unavoidable than accidental.

A Worker's Paradise?

Marxian apologists will point out that the communist world has achieved some remarkable successes in the social realm. The former Soviet Union, for example, has much less homelessness, malnutrition, illiteracy, and drug addiction than does the United States—a country of vastly greater economic prowess. There is indeed some truth here, and the social failures of the United States should be considered a national shame. The comparison, however, is invidious in that it singles out the industrial capitalist nation with the worst record on social issues. No other wealthy capitalist state, for example, lacks a national health care system. If one were to contrast Japan or Sweden with the USSR—let alone with Romania—capitalism would come out ahead on virtually every social issue as well.

The social failure of marxian socialism is probably best illustrated by examining the working and living standards of its own laborers—the supposed beneficiaries of the whole system. Simply put, socialist workers lived in penury when compared to their counterparts in industrial capitalism. Polish steelworkers, for example, could hope to earn roughly the equivalent of $100 a month; if one were to factor in the loss of time entailed in queuing, their remuneration would have to be reduced still further. But such deprivation is utterly mild when contrasted with the lot of Soviet coal miners—men who labored under such appalling conditions that their average longevity was a mere forty-seven years (*The Economist*, "Dark Satanic Mills," October 13, 1990, p. 56). Indeed, industrial safety standards have been virtually nonexistent through much of the Eastern block. Because of this failing, up to 80 percent of Polish steel workers were disabled and thus forced to retire early (Fischoff 1991:14).

According to marxist ideology, these Polish and Soviet workers were not exploited—even if their political leaders and party bosses were able to live in aristocratic splendor. ("Exploitation," one will recall, is defined in terms of the surplus extraction that occurs only under a capitalist mode of production.) Such reasoning, evidently, held little appeal for the Polish and Russian proletariat; despite the long years in which it has held absolute political, social, and cultural mastery, marxism was never able to achieve intellectual hegemony in eastern Europe. What seems inevitable now is the collapse of communism, not capitalism.

The Contradictory Success of Capitalism

The thesis that capitalism is destined to fail from its own internal contradictions is a bit threadbare these days. The present era is one of capitalism regnant, visible in the fall of communism in Eastern Europe, in the renouncing of marxism by sundry African regimes, and in the spectacular success of East Asian capitalist economies. If we are to begin addressing our environmental and social problems we must first come to grips with this fundamental reality.

True believers, however, persist in maintaining that capitalism's success is merely a mirage. Just as Christian fundamentalists still believe that the second coming is nigh, so too marxists stalwarts continue to see the collapse of Japan and the West not only as inevitable but as due very soon indeed. Thus, they proclaim, we are now in the age of "late capitalism." To illustrate the system's impending demise they point to the abundant signs of economic and social decay in the United States—the same signals that many fundamentalist Christians believe prove that Jesus is about to begin his descent.

Several fundamental errors, however, tarnish the crisis and decay thesis. Most importantly, radicals of all stripes consistently overestimate the signs of doom. To appreciate this we can play two schools of extremism against each other and in so doing arrive at a reasonable middle ground. The anti-environmentalist ideologue Ben Wattenburg (1984), on the one hand, argues that all measures of social well-being actually show signs of vast and continual improvement; we are misled by the apocalyptic prophets, he claims, because we see the past through rose-colored glasses. The eco-marxist James O'Connor (1987), on the other hand, sees only decay wherever he looks. Neither view is particularly instructive, and both would prove paralyzing if taken at face value. We should listen to both Pollyanna and Cassandra, but we would be foolish to accept either as offering accurate assessments or clear prophesies.

In a backhanded and unintended manner, the thesis of inevitable capitalist decay is actually belied in the writings of many contemporary marxists. Such scholars consistently and rightfully point to the damages caused by the Reagan and Bush administrations' social policies. James O'Connor, for example, argues that Reaganomics required us to sacrifice our "dreams of an equitable and just society" (1987:39). This sentiment implies, however, that the recently demolished social programs previously enacted by the Democratic Party were bringing justice and equity to capitalistic American society. Yet if marxism tells us anything it is that justice and equity are *absolutely* impossible under capitalism. Here we encounter a great intellectual game of "cake eating and having." When social progress is made within a capitalist society it is ignored or dismissed as chimerical; when social regression occurs it is highlighted as very real indeed—even if it entails nothing but the dismantling of programs previously denied as unreal. Such sophistry does indeed allow one to argue that capitalist society will only ratchet ever downward into more brutal forms of injustice and exploitation.

The thesis that capitalism is in inevitable decline is also parochial. The unstated assumption is that capitalism is congruent with the West, if not simply with the United States. Signs of decay in America are thus heralded as foretelling the decline of capitalism in general. Here marxists simply don the same blinders that virtually every American enjoys wearing. We find great comfort in believing that the United States is still the world's dominant country simply by virtue of its sizable gross national product (GNP) and formidable military. But America's present lead over Japan reflects not the two nations' internal economic dynamics, but rather America's tremendous head start, its larger population, and, to a lesser extent, its greater wealth in resources and land. On virtually any measure of economic transformation, be it industrial, technological, or financial, Japan is either ahead or closing in quickly.

The great French historian Fernand Braudel (1984) has shown that the world capitalist system is almost always centered in a single city. No city, however, has been able to retain dominance for long; capitalism is simply too competitive. Thus Venice yielded to Antwerp, Antwerp to Genoa, Genoa to Amsterdam, Amsterdam to London, and London to New York. And now New York is falling to Tokyo. As each former center, and its surrounding country, loses primacy, relative decay will be inevitable. In the rising star, however, such signs of decrepitude should be absent.

The essential question thus is whether Japan shows the signs of social and economic collapse that eco-marxists perceive in the United States. It is difficult indeed to argue that it does. Economic growth and productivity increases may have slowed down a bit, but they are healthy year after year; sundry economic shocks may provoke fear, but they are always contained after short periods. Inflation and unemployment levels are minuscule by Western standards, violent crime is rare, drug addiction is scarcely a problem, child abuse is almost unknown, and homelessness is nearly nonexistent. And whereas the wage gap between high school and college graduates is increasing in the United States, in Japan it is decreasing (Reich 1991:206). Not that Japan is a paradise; if one looks closely enough many social evils are readily apparent. Japanese society is, for example, shockingly bigoted and its progress toward women's rights is woefully retarded. But even here halting improvements are being made.

The belief that capitalism is marching ever onward to its inevitable decline is little more than darkly wishful thinking—wishful because marxists believe that true justice can be instituted only when capitalism expires, but darkly because marxists well know that the poor suffer disproportionally in times of chaos and crisis. The unfortunate truth is that in the United States the poor are being immiserated while the beneficiaries of our increasingly second-rate capitalist system grow richer and more numerous every year. This is a sign, however, not of doom but rather of the political failure of liberalism and of the relative slippage of the American economy within the global capitalist system.

Marxist Contradictions
Contemporary marxist academics shield themselves from many of the problems explored above by leaving their own ultimate aims unstated. Marxism's own traditions unambiguously uphold violent revolution as the only way to usher in socialism, although the numerous democratic communist parties of Western Europe show that many marxists evidently now consider a democratic "gradualist path" possible. American academic marxists, however, are seldom inclined to reveal their own larger designs. One can only assume that many admire Lenin and Trotsky not merely for their scholarly works but also for their activities in the Russian Revolution. The "scholar-revolutionary" seems to be an irresistibly romantic figure for academics humiliated by their drudge-like popular image. But this is never mentioned in polite company; theirs' is indeed a hidden agenda.

A more immediate question is how academic marxists view their own roles within capitalist society. The obvious answer is as intellectual workers whose job it is to lift the veils of mystification and thus allow others to comprehend the reality of their own oppression. But they communicate these truths largely to each other; with a few notable exceptions, marxist works are so theoretically heavy and jargon laden as to be completely inaccessible to the proletariat. Academic marxists do reach a wider audience through teaching, but most of their students are destined for the managerial rather than the working class. This is especially true for those employed by elite universities, institutions to which most young marxists seemingly aspire. But marxist theory explicitly holds that nothing revolutionary can come from the bourgeoisie, at least once it has triumphed over the aristocracy. The most one could hope to accomplish would be to subvert the minds of budding managers, lawyers, and engineers, thereby making them less effective capitalist functionaries. Once students leave academia, however, few retain the subversive tendencies they may have acquired from their marxist professors.

But no matter how one hopes to achieve it, the revolutionary transformation of capitalist society remains the centerpiece of marxist philosophy, the driving force behind the entire effort. Without the prospect of a socialist future, the voluminous marxian critiques of capitalism lose most of their power and much of their relevance.

Yet a successful marxian transformation, be it evolutionary or revolutionary, hardly seems likely within the United States. The evolutionary path is moribund; socialist parties never achieve more than a percentage point or two in any election, except in a few errant university towns like Berkeley and Santa Cruz, California—or in Vermont. So too the chances of a revolution in the near future, as most marxists fully recognize, are nil. But despite such dismal prospects, marxists cling to the hope that in the event of a severe socioeconomic trauma, success might be theirs'. Therefore, they strive to strengthen their position within academia and to build linkages with social movements (such as environmentalism and feminism) until a more favorable political environment emerges around them.

While an explosive socioeconomic crisis in the near term is hardly likely, the possibility certainly cannot be dismissed. Capitalism is an inherently unstable economic system, and periodic crises of some magnitude are inevitable. An outbreak of jingoistic economic nationalism throughout the world, moreover, could quickly result in virtual economic collapse. Under such circumstances we could indeed enter an epoch of revolutionary social turmoil. Yet I believe that there are good reasons to believe that the victors in such a struggle would be radicals not of the left but rather of the right.

The extreme left, for all its intellectual strength, notably lacks the kind of power necessary to emerge victorious from a real revolution. A few old street radicals may still retain their militant ethos, but today's college professors and their graduate students, the core marxist contingent, would be ineffective. The radical right, on the other hand, would present a very real threat. Populist right-wing paramilitary groups are well armed and well trained, while establishment-minded fascists probably have links with the American military, wherein lies the greatest concentration of destructive power this planet knows. Should a crisis strike so savagely as to splinter the American center and its political institutions, we could well experience a revolutionary movement similar to that of Germany in the 1930s.

Marxists, however, would likely counter this argument by citing the several cases of successful socialist revolutions. Successful though they were, none makes a compelling analogue. First, no marxist revolution has ever come close to occurring in an advanced capitalist nation. Triumphant leftist revolutions have only taken place in economically backward countries, and generally only after an unrelated war had demoralized the old guard. More importantly, as Hamerow (1990) clearly shows, all successful marxian revolutions have relied on the strategic cooperation of the bourgeoisie against the aristocracy; only after the old regime is toppled are the fractionated moderates cut out of power. Considering the fate that has generally befallen them under such circumstances, it is unlikely that the business classes—even in the world's more feudal countries—would again be tempted by the promises of a mixed economy offered to them by would-be leftist revolutionaries. Except perhaps in El Salvador and Peru, contemporary marxist revolutionary movements are irritants to the ruling elites rather than real threats.

In contemplating the likely future of a revolutionary United States, we encounter the ultimate paradox of contemporary marxism: the unintended collusion of the radical left and the radical right. Even during periods of normality, the opposing ends of the political spectrum feed strongly on each other—in sardonic fashion, they are each other's best allies. The marxian left is extraordinarily frightening to the vast majority of the populace, and the stronger it becomes, the more seductive the propaganda of the radical right grows. The equation can also be reversed; leftist rhetoric draws its real power in opposition to the radical right, not the accommodating center. With every KKK outrage, with every atrocity committed by the Los Angeles Police Department, the marxian message grows ever more convincing to horrified progressives. The broad center of responsible conservatives, moderates, and liberals may attempt to remain dispassionate and to refute both extremes, but in a deteriorating political environment, marked by inflamed passions, such a stance will seem to many increasingly inadequate.

If, in the event of extraordinary crisis, the center does fold, I must conclude that most Americans would follow the far right rather than the far left. American society has simply been too prosperous, and the majority of its citizens too accustomed to owning property, to be willing to risk everything on a communist experiment. Alexander Cockburn of *The Nation* has repeatedly pleaded with liberals not be afraid to endorse socialism—a fine position indeed if one would like to see reactionaries gain uncontested power throughout the United States. If truly concerned about social justice and environmental protection, I would counter, liberals should not be afraid first to embrace, and then seek to reform, capitalism.

American marxism is thus intrinsically paradoxical; not only is it self-defeating, but it actually reinforces (in a perverse antidialectic) its own antithesis. And if that antithesis ever gains power, it will not merely retain the status quo, but rather pull society fiercely backward, leading it into a truly nightmarish world.

But a critique of marxism, now matter how powerful it may be, will fail to impress the majority of eco-radicals, individuals who have never accepted more than Marx's basic arguments against capitalism. In order to further the cause of Promethean environmentalism it is necessary to show how capitalism can be transformed into an ecologically benign economic system.

■ Guided Capitalism and the Environment

The Amorality of Capitalism

Radicals are correct in arguing that capitalism is based on naked self-interest; this has been abundantly clear ever since Adam Smith. Capitalism is an amoral, although not necessarily an immoral, system. Morality must be derived from an exterior source, and an unguided capitalist economy will tend toward great brutality. Such an ethical backdrop has often been lacking, and countless capitalist firms have perpetuated horrendous evils. Mid-century Germany provides the ultimate example. While nazi ideology was virulently anticapitalist, most German capitalists were happy to cooperate with Hitler once it was clear that he did not really intend to dismantle their corporations.[3]

Leftist academics, however, overemphasize capitalism's lack of morality by confusing the motivation of the individual firm with the requirements of the system in general. Rapacious strategies may benefit a given company, but they undermine the larger structure of capitalism. Such confusion is especially evident in studies of South African racial exploitation, a field that has come to be almost entirely marxian in orientation (Murray 1988). By briefly exploring the failings of the main marxist interpretation of apartheid we can better understand how the interests of the individual firm diverge wildly from the interests of capitalism in general.

Many prominent leftist scholars have argued vigorously that apartheid exists because it is functional for South African mining companies, especially by ensuring them a steady supply of cheap labor (for example, Wolpe 1972). Commonly overlooked, however, is the fact that members of the South African capitalist class have, more often then not, opposed apartheid. As Bromberger and Hughes (1987:204) write: "[I]t is not at all clear that capitalist interests were predominant in the creation of South Africa's modern racial order. Some were *not* significantly involved; and over time capitalist support for racial discrimination and controls has diminished and turned, in most sectors, into opposition." Support for apartheid has rather come mainly from the white working class, the white farmers, and the privileged Afrikaaner bureaucracy (see Parnell 1991). As Lipton (1985:370) concludes: "[South African] development since Union does not support the thesis that the state was the instrument of capital. The interests of the economically dominant mining and urban capitalists were often overridden when they were in conflict with those of white labour or the bureaucracy or of economically weaker agricultural capital. The key to this lay partly in the political system, partly in the nature of political mobilization, which was along ethnic, not class, lines. Afrikaaner nationalism is the most striking example of this, and it poses a severe problem for Marxist analysis."

Business leaders have opposed apartheid not because of their magnanimity, but rather because discrimination is in many respects highly disfunctional for the economy. Many South African companies have long suffered from shortages of skilled labor, yet they have been politically prevented from tapping a huge segment of the populace for such positions. As a result, wages for white workers have been far greater than the market would dictate, a situation hardly advantageous for capital. Even more importantly, the fact that so many people have been reduced to dire poverty by political edict greatly reduces the internal South African market, which in turn undercuts the potential profitability of consumer-goods firms. The underdevelopment of the consumer economy, in turn, severely hampers the country's overall economic performance.

The same underlying patterns may be seen, albeit in weaker form, in the United States. It was, of course, the capitalistic Republican Party that dismantled slavery; until relatively recent times the Democratic Party of workers and farmers formed the bulwark of discrimination. As a system, capitalism thrives on equality of opportunity. Efficient corporations welcome talented individuals from all social ranks into their middle and upper echelons—so long as they are adept at making profits. Thus the editor of *Fortune* magazine tells us that "One of America's great competitive weapons is that we are far ahead of the Japanese and most other foreign competitors in at last beginning to admit women to positions of real power" (July 30, 1990, p. 4). Of course, individual capitalists can be as bigoted as any one else, and many are blind to the general requirements of the system as a whole. And so too, equality of opportunity must never be confused with social equity, as those individuals lacking the demanded skills and motivation will always be poorly rewarded by the rational corporation.

Although capitalism, in the end, precludes economic equality, it does suffer if wage differentials grow too great, as we have already noted in the case of South Africa. As many marxist scholars now recognize, low wages across the board translate into minimal purchasing power, which is hardly advantageous for a capitalist machine often desperate to find markets for its abundant goods. Thus, in the virtuous capitalistic spiral of "Fordism" (Scott and Storper 1986), productivity gains have been partially shared with workers in the form of higher wages; the aggregate result being a prosperous working class and a healthy economy.

It is also essential to recognize that state-mandated social and environmental regulations can actually aid the capitalist system, even if they do burden individual firms. A capitalist society cannot long persist if individually rapacious companies are allowed to destroy their workers or demolish the environment. Despite marxian—and reactionary—claims to the contrary, environmental regulation has contributed little to our economic slowdown (Leonard 1988:57); many experts would go so far as to argue that it functions in the long run to enhance national competitiveness (for example, Porter 1990). Similarly, many American corporate directors are realizing that Japan's socialized medical system gives its firms a profound advantage in competing against American companies suffering under heavy health insurance burdens. Nor is it coincidental that the most successful capitalist economies of the past two decades— Japan, South Korea, and Taiwan—are distinguished by their relatively equitable distributions of wealth, whereas those countries with the greatest gaps between the rich and the poor, such as Brazil, have faltered mightily (see chapter six). It is for this reason that conservative proponents of capitalism would be well advised to reexamine the recent socioeconomic history of the United States.

American Economic Evolution

The heyday of American capitalism was the 1950s and 1960s, an era of steady economic growth, healthy corporate profits, increasing labor productivity, general prosperity, and decreasing pay gaps between blue- and white-collar workers. Presiding over this vibrant economy was a series of moderate Democratic and Republican administrations that fully accepted the progressive tax code and social security measures implemented in the 1930s. Most importantly, loose agreements between large industrial companies and labor unions functioned to distribute wealth generated by productivity gains among workers, managers, and owners.

Postwar prosperity did not, of course, usher in a capitalist utopia. Large segments of the populace remained impoverished, often excluded from decent jobs by naked discrimination, while accelerating environmental destruction was virtually ignored. By the mid 1960s, however, pressure was mounting to address the problems of civil rights, poverty, and environmental destruction. For a while it appeared that a new, more encompassing form of liberalism might triumph, but this was not to be the case. Tragically, many of those fighting against injustice came to embrace not liberal reform but rather a left-radical philosophy that sought to bring down the entire system. This intransigent leftward lurch, coupled with a stagnant economy in the 1970s, resulted in increasing political polarization, culminating in the triumph of Reaganism in 1980.

Under the Reagan administration, American capitalism enjoyed only moderate success. The wealthy stratum of the populace grew richer and larger, but the poor layer also grew larger—and poorer. Environmental standards were gutted, allowing many irresponsible firms to reap profits far larger than deserved. Radical Republicans see the era's economic expansion as proof that social programs and environmental protection thwart capitalist prosperity. Much evidence, however, suggests that they are gravely mistaken. The American economy in the 1980s was a pale counterpart to that of the 1950s and 1960s, a fact immediately confirmed by contrasting figures on productivity gains. Moreover, the prosperity of the 1980s was in part based on money borrowed from Japan.

One cannot, however, lay all of the blame for recent economic polarization at the feet of the Republican presidency. The continuing evolution of capitalism has itself made the comfortable world of Fordism obsolete. As many marxist scholars correctly argue, we have now entered an age of "post-Fordism." Firms are now not only increasingly unconstrained by national boundaries, but they are even freeing themselves from the barriers separating them from other corporations. As discussed in chapter three, the world of capitalism is now characterized by flexible networks of strategic alliances, subcontracting linkages, licensing arrangements, and so on. As the previously monolithic corporate structure begins to dissolve, Reich (1991) cogently argues, economic elites lose their need to maintain old agreements with production workers; cheap labor, after all, can now be easily obtained in poor countries. Similarly, if environmental regulations become too stiff, executives can relocate their plants in foreign havens unmindful of ecological costs.

We obviously cannot return to the stable and comfortable world of the 1950s. But we by no means must fatalistically assent to the growing polarization of American society. To resist this insidious trend we must first recognize that it is contrary to our national interests. Despite the growing internationalization of capitalism, the national economy—with due respect to Reich—is not going to evaporate. And any national economy that excludes large segments of its population from meaningful participation necessarily impoverishes itself. Reversing the tide of economic stratification will, however, require new foundations for a moderate liberalism. The essential task is not to replace capitalism, but rather to make the social and infrastructural investments necessary to ensure that the entire population can participate within it at reasonably high levels. In order to do this, we must first come to appreciate the powers and problems of competition.

The Powers of Competition
In emphasizing the atomizing or individualizing side of capitalism, eco-radicals blind themselves to its opposing social aspects, thus ignoring a tradition of thought dating back to Smith and Hume (Novak 1990). This misconception is nowhere more evident than in discussions of competition. Especially among anarchists, a simplistic Spencerian view prevails, one viewing competition as a desperate struggle for survival, the winner taking all and the loser perishing. To most radical environmentalists, competition is intrinsically antisocial, deadly to cooperation, and contrary to the principles of ecology. Even the relatively moderate eco-philosopher Kassiola (1990) centers many if not most of his arguments around the inherent destructiveness of competition in all of its forms.

But as anyone who has ever played sports knows full well, competition usually *creates* strong social bonds. Camaraderie not only links team members, but it can even develop between opposing contingents. Such bonding between competitors is most clearly evident in nonteam sports; people generally play tennis or racquetball with their closest friends, not their most bitter enemies. Noncompetitive sports, on the other hand, not only fail to bring out peak performances, but they seldom prove satisfying. When I was an undergraduate at the University of California at Santa Cruz, where marxist, primitivist, and anarchist philosophies prevailed, many students declared that they would no longer play games such as racquetball in competitive manner. Instead they would merely bat the ball around for a while, and in so doing spare the egos of the less-skilled players. In short time, however, most of these caring persons ceased playing altogether.

Economic competition can be analyzed in much the same terms. Here too, "cooperation and competition are opposite sides of the same coin" (Dertouzos et al. 1989:94). Team spirit is generated within successful firms, and even between competing firms a healthy spirit of friendly competition sometimes prevails. Cooperation, moreover, is essential between firms that supply and purchase from each other; not surprisingly, this is an area in which Japan has excelled. Any given company can, of course, display bad, even criminal, sportsmanship. Thus an external referee—the state—is necessary, just as it is in sports. Corrupt or incompetent state officials can still fail at their duty, but this calls for greater vigilance, not repudiation of the system.[4]

Radicals of both the right and the left fundamentally reject the team metaphor of corporate organization. Individuals on both extremes prefer to picture society as internally bifurcated into the inherently antagonistic camps of capitalists and managers, on the one hand, and workers on the other. Such natural enemies are not considered capable of striving for common goals; instead, they are viewed as consistently acting at cross purposes. Radical right-wingers of the old school view workers as uneducable riffraff of dubious moral qualities, individuals who no more form part of the team than do the janitors who clean up a ballpark after a game. Radical leftists, for their part, see capitalists and managers as predatory exploiters who will do everything they can to keep wages at a level just adequate to sustain life. Both groups thus believe that it is essential to place all power on one side of this unbridgeable divide. Rightists want to discipline workers firmly, while leftists would like to eliminate capitalists entirely.

Yet the belief that workers and capitalists have fundamentally divergent interests is no longer reasonable (Thurow 1985). If a firm fails, all fail with it; if it succeeds, all can prosper together. A partnership model, one that regards workers, managers, and investors as striving in team-like fashion, is potentially advantageous to both capital and labor. All parties contribute essential services both to the individual firm and to the national economy. As Reich (1987) demonstrates, it is "collective entrepreneurialism" that ultimately determines whether or not firms succeed.[5]

Leftists scoff at such a corporatist model. And well they might, as it is deeply threatening to their own agenda. If laborers and managers really begin to work together, if wages are universally replaced by salaries, and if shop-floor workers are increasingly rewarded through profit-sharing and stock-option plans, the very distinction between the bourgeoisie and the proletariat begins to vanish. Of course, one should never expect a corporatist utopia; different factions will still fight to maximize their own benefits, and those individuals possessing more education, skill, or motivation will continue to be more richly rewarded than others, generating perennial ill-will. But in relative terms, an economy based on cooperative competition is potentially far more powerful *and* equitable than one based on contention alone.

To ensure efficiency, the economy must be divided into a multitude of competing firms. As Michael Porter (1990) so well shows, both capitalist firms and national economies thrive only in the atmosphere of challenge and adversity that fierce competition provides. But at a higher level, the entire country must be seen as forming a single "team"; a group of hundreds of millions of individuals sharing common interests in prosperity, justice, and environmental protection. The role of the government is thus not only to intercede between firms to ensure healthy competition, but also to encourage economic growth and to provide public goods—including environmental protection. How active the government's economic role should be is a matter of much needed debate, but as liberal political economists amply demonstrate, America already has an "industrial policy" (Thurow 1985, Rosecrance 1990, Reich 1991). The problem, however, is that this policy has been vague, inefficient, and altogether too closely tied to the military.

This same argument can also be extended to the global scale. National economies, despite their growing interdependencies, will continue to compete against each other. If competition is fair, overall efficiency will increase, generating greater global prosperity. International mechanisms should, however, be strengthened both to ensure fair play and to abet those countries recently devastated by imperialism and now struggling to industrialize and join the world economy. And ultimately, we should regard all of humanity as members of a single team; we may not face extraterrestrial competitors, but we do share the same home. If that home is destroyed, as it may well be, we will all find ourselves miserable losers indeed.

Capitalism and the Environment
Innumerable capitalist firms have indeed inflicted tremendous damage on the environment in pursuit of their own interests. But such degradation, however much it may benefit the companies responsible for it, is not without economic costs. In the most direct example, a factory can so pollute a stream that its waters may no longer be usable by another industrial plant located downstream. Damage is more often diffuse and difficult to assess, but whenever pollution results in reduced crop yields, increased incidents of disease, or loss of recreational opportunities, to give just a few examples, the economy as a whole suffers substantial damage.

Such costs that can be ignored by the firms responsible for them and instead passed on to society at large are defined as "negative externalities." Many environmental economists regard negative externalities not as unfortunate side effects of the market system, but rather as incidents of market failure. In an efficiently operating market economy, an outside arbiter (some arm of the state) will require the damage-causing firm to pay for the costs it inflicts, thus "internalizing the externalities." Much research conducted by environmental economists assesses the economic damage resulting from environmental degradation and seeks to determine how polluting firms might be made financially responsible.

Many environmental economists also seek to reform the conventional method of setting the discount rate (by which future benefits are given a reduced value in comparison to benefits that can be immediately realized). If the stability of ecosystems is to be taken into account, some argue, a significant lowering of the discount rate may be both ecologically and economically wise (see *The Economist*, March 23, 1991, p. 73). Other concerned scholars, however, believe that there are more appropriate mechanisms for protecting posterity (see Pearce et al. 1990). At any rate, it is essential to recognize that businesses usually discount the future much less than do individual consumers, most of whom resist buying energy-saving appliances even when it is clearly in their own best financial interests (Bevington and Rosenfeld 1990:77). It is also important to realize that capitalist firms do not have uniformly short time horizons. Indeed, one of the main components of Japanese economic success seems to be the abnormally long time spans employed by its corporate managers. Future work by environmental economists may thus help us to reorient our economic time scale, and in so doing benefit both the environment and the economy.

Despite its vigorous advance in recent years, environmental economics does not offer a panacea. Under certain circumstances a firm can deforest a landscape or even exterminate a species without significantly harming other economic endeavors. Environmental economists may argue that economic damage is still inflicted on society at large, since people can no longer enjoy the forest or the extinct species. Therefore, they strive to determine the precise monetary value that individuals are willing to place on different aspects of nature (for example, Willis and Benson 1988). This is sometimes a relatively straightforward exercise, as in the case of ascertaining the market value of access to a national park. More ambitious scholars, however, seek to discover the value people place on the mere existence of natural features that they never intend to experience directly. Such studies can reach the point of absurdity, assuming as they do that everything can be accorded a specific price.

In the end, environmental economics offers powerful tools for combating pollution, but its utility for ecosystem preservation remains limited. Here radical greens do have a valid point; conventional economic analysis rests on the unconscionable assumptions that human beings are the measure of all things and that other species have no value except insofar as humanity might use or more passively enjoy them. But the deep ecological alternative, a relativism run wild that accords blue whales and smallpox viruses the same intrinsic worth, is just as dangerous. Ultimately, we need both an overriding ethical system that accords intrinsic worth to nonhumans—without insisting on pan-species egalitarianism—and an efficient means of allocating scarce resources.

Market Mechanisms and Environmental Protection

The economic approach to pollution control deserves the concerted attention of liberal and moderate environmentalists. The recently dominant nonmarket regulatory approach encourages the use of inefficient abatement methods. The EPA has usually set maximum limits for the discharge of specific pollutants, threatening firms that exceed them with legal action. Indeed, it often goes so far as to mandate the use of specific control technologies. Such a system, virtually all economists agree, is inherently inefficient; firms have no incentive to reduce their effluents below the set limits, and they may even be prevented from employing the most cost-effective control systems. Moreover, companies generally have little to fear if they exceed mandated discharge levels. Excess pollution may be penalized, but only a few small fines are ever levied. It is often in a firm's best interests to continue polluting at criminal levels, to hire a team of skillful lawyers, and simply to pay whatever meager fines may be assessed.

Economic research indicates that a market-based approach to pollution control could accomplish the same results as the regulatory system at a much reduced level of expenditures; a savings rate of some 30 percent is a typical figure (Blinder 1987:153). Alternatively, society could continue to devote the same amount of money to pollution abatement as it currently does, yet enjoy a much cleaner environment. With the advance of the recent Clean Air Act, official policy has indeed swung sharply in the direction of market mechanisms. So compelling is the case, according to *Forbes* magazine, that many mainstream environmental groups are developing a newfound respect for Adam Smith (see "Shaking the Invisible Hand," April 1, 1991, p. 64).

Market mechanisms for pollution control fall into two basic categories. The first is a system of effluent taxes. Under such an arrangement, firms are allowed unlimited discharges, but are forced to pay dearly for the privilege. The second system is one of tradable permits; a specific pollution target is set by the state, and then permits to emit waste up to that level are auctioned off. Firms are not allowed to pollute unless they hold the necessary permit. They can, however, buy and sell discharge rights among themselves, with the value of the permits fluctuating according to market conditions.

Both effluent taxes and marketable pollution permits provide powerful incentives for firms to reduce their discharges. Under the former system, a factory might have the theoretical right to spew out unlimited quantities of waste, but economic considerations would make this impossible; as discharges increase, the tax burden would eventually become unbearable. If, on the other hand, a firm can continually reduce its waste stream, its tax burden will progressively decrease. Pollution will never be eliminated (at some point the marginal effort to reduce discharges further will not be cost effective), but then neither does the regulatory approach promise total cleanliness. The permit system offers similar incentives; a permit-holding firm will strive to reduce its effluents so it can sell its pollution rights and thereby gain additional revenue.

Under either market-based approach, pollution control measures would be implemented first where reduction is most efficiently accomplished. A certain factory might be able to reduce easily its discharges to a small fraction of their previous level, but another may have little option but to purchase more permits or pay higher taxes. The firm owning the second factory, however, still has a strong financial incentive to purchase or devise new, more efficient, control processes. In a regulatory system, on the other hand, all firms are often required to reduce their discharges to exactly the same level. This means that some factories may continue to spew out pollutants that they could easily eliminate, while others may be forced to shut down immediately, sending thousands of workers into unemployment (see Blinder 1987).

Radical environmentalists quickly dismiss the use of market incentives to control and reduce industrial waste. They claim that such a system only legitimates pollution, as if fouling the environment were some kind of right. Much better, they argue, to treat pollution as a criminal matter. But under the regulatory system, firms still have the right to pollute—so long as they do not exceed a certain arbitrary limit. More importantly, the regulatory approach provides no incentive for firms to reduce their discharges below their rightful allotment.

Considering the clear environmental advantages of the judicious use of market mechanisms, continued opposition indicates that some eco-radicals may have more hostility toward capitalism than concern for nature. If an incentive system can achieve greater pollution reduction than a regulatory approach at the same level of social expenditure, one would expect all environmentalists to support it. If, however, the market is dismissed beforehand as invidious while efficiency itself is suspect as the handmaiden of a nasty capitalist rationality, then the environment will be forced to bear the burden of ideologically mandated degradation.

Interestingly, the radical right-wing "free-market environmentalism" propounded by writers such as Anderson and Leal (1991) is also hostile to the use of market mechanisms. According to these authors, effluent taxes are unacceptable because they must be established by the state. Rather, they favor a system in which all environmental resources are privatized so that individual holders can protect the environment by suing those who degrade their own personal domains. This vision is not merely anticommunal but actually antisocial. Fearing that sightseers may get a free-ride, Anderson and Leal (1991:20) seriously advocate fencing off areas of natural beauty so that private owners could charge others for the privilege of looking at them.

Anderson and Leal, as radical free-marketers, prefer to let the market work its magic without guidance—except by the process of interminable litigation. (As Robert Kuttner argues, "The flip side of rugged individualism and weak government is endless lawsuits" [*Business Week*, June 3, 1991, p. 16].) Indeed, they go so far as to argue that the free market, with its give and take, is structured almost exactly like an ecosystem (1991:5).

Contrarily, the environmental leftist Michael Redclift (1988:55) advises us to shun market mechanisms because "the properties of ecological systems run counter to those of . . . neo-classical economics." Neither view, however, is at all instructive, as both involve a theoretically naive reification of conceptual models of both the environment and the economy.

Regardless of extremist fantasies, we can expect that once capitalist energies begin to be harnessed to environmental protection, a virtuous spiral will begin to develop. Several American companies, for example, have already pledged to reduce their discharges well below current legal limits. Such firms foresee stricter regulations in the future, and they are not unmindful of the desirability of maintaining good public relations (which, contrary to the green radicals, should be hailed as a powerful force for reform, not disparaged as mere window dressing). Moreover, in learning how to reduce their own effluent streams, such companies will devise new control mechanisms and strategies that they may be able to sell profitably to environmentally retarded firms in a more ecologically aware future world. Leading-edge corporations may eventually have a vested interest in the enactment of stricter pollution control legislation. As *The Economist* prophesies: "The greenest companies will therefore try to ensure that government policies set environmental standards at levels that they can match but their competitors cannot. The greenest governments will see such companies as potential allies, and will try to promote policies that foster investment in environmentally friendly technologies" ("Survey of Industry and the Environment," September 8, 1990, p. 20).

Economic Growth and Pollution Abatement

But no matter how efficiently it is accomplished, pollution control will be costly in the short run. The cleaner we wish our environment to be, the more money we will be forced to spend. How then should society pay the bill? The radical approach, to force all costs on industry, is tempting but ultimately unworkable. It is unlikely indeed that corporations would meet such requirements solely by reducing their dividends or executive salaries. More likely, they would pass their additional costs to consumers through higher prices or to workers through lower wages; alternatively, they might simply go bankrupt. Ultimately, the economy as a whole will be forced to bear the costs of environmental protection. Prices will be higher, while wages, salaries, and dividends will be *temporarily* lower than they would in a society unconcerned with the natural environment. (In an environmentally oblivious society, however, the destruction of ecological systems will *eventually* undercut all economic gains.)

But environmental protection need not result in economic decline. We can pay for pollution reduction without sacrificing business health or consumer prosperity, but only so long as we have a vigorously expanding economy. In a growing economy, extra increments of wealth can be channeled into environmental protection without being detracted from existing endeavors; in a steady-state (or "zero-sum") economy, in contrast, resources for pollution control must be diverted, painfully, from other economic spheres.

Radical greens, of course, would just as soon see huge sectors of the economy starved out of existence. Yet they consistently ignore the repercussions of economic stagnation on the politics of environmental protection. In a buoyant economy, individuals are often willing to devote large sums to the public (or "natural") good, knowing that such altruism will not jeopardize their own standings. In a stagnant economy, on the other hand, the electorate often grows more cautious, seeing personal threats in every potential public outlay. Thus, popular support for environmental protection in the United States mushroomed in the robust 1960s, declined sharply in the anemic 1970s and early 1980s, and then rose again with recovery in the mid-1980s.

The notion that economic growth may benefit the environment is anathema to the radical greens. Their foundational belief—that expansion will ultimately destroy the planet—is, however, growing more untenable year by year. Recent economic history demonstrates that an economy can expand while significantly reducing its consumption of both energy and key resources. "Since the oil embargo of 1973, energy intensity—the amount of energy required to produce a dollar of U.S. gross national product—has fallen by 28 percent" (Fickett, Gellings, and Lovins 1990:65). Similarly, the growing American economy has been continually reducing its dependency on numerous mineral resources. Some two decades ago, the Club of Rome (Meadows et al. 1972) predicted that copper shortages could soon spell the end of civilization, a view that now appears quaint as copper telecommunications lines yield to fiber optic cables made ultimately from sand. As Piers Blaikie (1989:130) tersely writes, the limits-to-growth thesis has been subjected to a "number of thorough debunkings."

In fact, as early as 1973 a group of environmentally concerned economists demonstrated clearly that the imperative was to reform rather than to end economic growth (Olson and Landsberg 1973). Several of these writers discerningly pointed to the dangers present in a no-growth economy, notably including a loss of freedom (McKean 1973) and the possibility that in "the stationary economy, unfortunately, investment in exploitation may pay better than in progress" (Boulding 1973:95). It is not at all coincidental that American liberals have consistently advocated economic expansion, whereas traditional conservatives have been far more concerned with stability (Kuttner 1991).

As a fitting epitaph to the exhausted idea of economic limits, one might inscribe the terms and the outcome of the Ehrlich-Simon wager of 1980 (Tierney 1990). In that year the ecologist Ehrlich bet that the prices of five key minerals would increase over the following decade as natural deposits were consumed; the economist Simon countered that prices would drop as substitutes were developed and new deposits discovered. When the price trends were tallied in 1990, not only did Simon come out ahead, but he would have triumphed even if the terms had not been indexed for inflation. In light of this and other evidence, I believe that we can now safely conclude that the future of advanced technology and of capitalism does not ride on the continued availability of tungsten or tin.

The "limits to growth" hypothesis is ultimately similar to Jeremy Rifkin's notion that we should expend as little energy as possible in order to forestall the eventual heat-death of the universe. Limits do exist for specific resources, but in the most important cases they are so remote as to be virtually meaningless. Using the same logic one could declare all human endeavors futile, seeing that the sun will eventually go supernova and consume everything. More importantly, environmentalists must come to understand that economic growth increasingly entails not the ever mounting consumption of energy and raw materials, but rather ever increasing value added—which as often as not is accomplished through miniaturization, partial dematerialization, and the breakdown of the very distinction between goods and services (*The Economist*, "Survey of Industry and the Environment," September 8, 1990, p. 25; see also Reich 1991).

But in rejecting the limits-to-growth thesis one is by no means obliged to accept Simon's competing cornucopian paradigm, a view marked by striking hostility to nonhuman life forms and by the celebration of ever increasing human numbers. The economy can continue to expand because it is based on value rather than mass; human beings, on the other hand, are unavoidably substantial. Not only are there firm limits to the human population the earth can support, but all environmentalists are obligated to strive to arrest the human tide well before such limits are approached. If we fail here, untold damage will be inflicted on all natural communities.

Fiscal Environmentalism

Although economic growth is environmentally desirable, one cannot assume that a healthy economy can be obtained simply by allowing the market to work its magic. To the contrary, social organization and governmental support and guidance are essential. In particular, we can only hope for environmental salvation if we are able to develop a long-range economic view that stresses investment, especially in human capital. To achieve this, major reforms in fiscal policy will probably be necessary.

As Paehlke (1989: chapter eight) has noted, there are surprising affinities between moderate environmentalism and supply side economics; both favor investment and both denounce unconstrained consumption. The fundamental difference, however, is that environmentalists vehemently reject the conventional supply side belief that investment capital can be augmented by lowering the tax burden of the rich while consumption should be limited by immiserating the poor. This position is both immoral and erroneous; as many observers have noted, the rich devoted most of their Reaganite tax gains to lavish consumption (Reich 1991: 265, 266). While we must never ignore demand, the further development of a *liberal* supply side economics (Dionne 1991:256) is highly desirable for both environmental and economic reasons.

Social justice alone demands fundamental income tax reform. The United States now employs a horrendously regressive taxation system, one that sometimes forces the poor to relinquish a higher proportion of their income to the government than the rich (provided one factors in social security payments). We could do much to rectify this situation merely be returning to the tax system—adjusting income brackets for inflation—employed during the days of Eisenhower. But if income tax reform is the immediate social need, long-term economic and environmental requirements call for a wholly new tax code, one that would encourage long-term investment. The central question is thus whether social equity and investment augmentation are necessarily in contradiction to each other.

Perhaps we should begin by questioning whether taxing income is the correct tactic in the first place. Lester Thurow (1985) argues, as a stanch Democrat, that both the corporate and individual income tax should be eliminated. The corporate tax, he claims, is merely passed along to consumers, workers, or investors, while income taxes in general discourage saving. Consumption taxes, like the VAT (value added tax), he argues, cause fewer economic distortions and could thus help transform our economy into one that emphasizes long-term growth (see also Dertouzos et al. 1989:145). Although consumption taxes are generally re-

garded as socially regressive, they could be manipulated to fall disproportionally on the shoulders of the rich. Indeed, some business leaders are
now advocating the creation of a socially progressive consumption tax
system—if accompanied by a lowering of tax pressures on productive investments (*Fortune*, April 22, 1991, p. 119). Such a system could also provide a number of environmental benefits. If consumer goods were taxed
in accordance with the degree of degradation entailed in their manufacture and disposal, for example, the environmental savings would be
immediate and profound (see also MacNeill et al. 1991:39).

A socially and environmentally oriented consumption tax system
would begin by acknowledging that the wealthy spend vast sums of
money on luxury objects, many of which entail substantial environmental damage. Stiff taxes on high-class automobiles, yachts, private airplanes, furs, foreign travel (especially to wealthy countries), and second
homes, to give just a few examples, would be inherently progressive as
well as environmentally salvaging. Considering how often luxury goods
are imported, such a fiscal policy would also improve the American trade
balance. And considering the amount of time and effort now wasted in
arcane income tax calculations, vast amounts of social energy could
be freed from tax deliberation and instead channeled into productive
activities.

Some necessities, it must be acknowledged, entail extraordinary environmental destruction. Gasoline is the prime example. To allow the
creation of an ecologically sane transport system, gasoline must be massively taxed—as it is now in Japan and Europe. But such a measure would
be regressive, as it would force large numbers of poor people to seek
alternative modes of conveyance, while sparing the wealthy any inconvenience.[6] Removing other tax burdens from low income groups and investing in public transport could help restore equity, although the latter
measure would be of little assistance to the rural poor. In the end, some
contradictions between social and ecological imperatives are probably
inescapable.

The Reinvention of Bourgeois Values

Perhaps the creation of an environmentally benign economic order calls
for a return to, or an invention of, a truly capitalistic ethos. Capital itself
must be regarded as virtually sacred—it represents nothing less than the
savings necessary to construct a more prosperous and less environmentally destructive future economy. Capital is deferred gratification writ
large. Just as a child must learn to delay the satisfaction of his or her
wants in order to become a responsible adult, so too the public must

learn to put away for the future. This imperative will only become more salient as we struggle to develop a new, environmentally benign economic base. As Ogden and Williams (1989) demonstrate, the transition to a PV-hydrogen energy system will entail extraordinary capital costs. Moreover, as Paul Magnusson shows, the United States is already "stumbling worst in technological areas where large capital investments are required" (*Business Week*, April 1, 1991, p. 27).

American society, however, has largely become one of spoiled economic children; individuals who would rather borrow than save, and who enjoy squandering their earnings on frivolous pursuits and meaningless status symbols. As Krugman (1990:47) demonstrates, the fall in the American savings rate is ultimately responsible for many of our economic ills, including our massive trade deficit. Unfortunately, this is not merely an American problem; savings rates have been shrinking in most industrialized countries in recent years, threatening numerous companies with capital shortages (see *The Economist*, "Survey of Capitalism," May 5, 1990, p. 6). Ironically, failure to save seems to be most strongly evident among the wealthy proponents of certain right-wing philosophies. As Robert Reich (1987:27–28) aptly reminds us, traditional conservatism "spoke of austerity and self-discipline," whereas the modern version "preaches austerity and discipline, to be sure, but with the crucial revision that the discipline is not for 'us' but for 'them'" (in other words, the poor). This sorry situation is ironically reinforced by certain radical capitalist enthusiasts who inform us that we should devise institutions "that firmly put the interest of the individual as *consumer* above his interest as *producer*" (Seldon 1990:204).

Environmentalists call for voluntary simplicity, and many look to "primal" peoples as exemplars in this regard. The concept is sound, but the example is faulty. Members of small-scale societies often live materially simple lives involuntarily; not uncommonly they covet the goods of the industrial world—witness Melanesian cargo cults. The best examples of voluntary simplicity may be found in capitalist societies, from groups such as the Calvinist burghers of the early modern period to the Japanese "salarymen" of the present. Although the sentiment is now weakening,[7] the Japanese—as a people—have made a commitment to live less prosperously in the present so that they might enjoy a more prosperous future. Similar patterns are visible in the newly industrialized countries of East Asia that have recently climbed out of poverty; all have successfully constrained consumption in order to stimulate investment (see Stallings 1990:72; Bradford 1990:41). Moreover, whereas extremists of both the left and the right tell us that capital can only be accumulated by the rich, the example of Taiwan shows clearly that accumulation can be a broad-based phenomenon.

If this line of reasoning is valid, one of our central failings is that the business class has been infected with the vanities of the aristocracy, a trend, admittedly, of long standing (Hobsbawm 1975:260–61). The failure of the American rich to invest their tax savings in the early 1980s is indicative of this sorry situation. In part it is a crisis of confidence; the fact that so many leaders of industry should feel compelled to advertise their positions with pathetic little trinkets like power watches shows only how insecure they really are. If anyone doubts this argument they should examine the extravagances of so many of our failed savings and loan institutions; one Miami firm boasted gilt sinks in its executive washrooms (whatever precious metals may have graced the executive toilet seats have not been disclosed). We can also witness the aristocratization of the bourgeoisie in the development of what has been felicitously called "casino" or "punter" capitalism—the practice of betting massively on momentary market fluctuation, thus forsaking long-term productive investment. The aristocracy has always adored gambling; members of the business class once preferred to invest. Just as aristocracies everywhere fell to the rising industrialists, so too nations entranced with windfall gains will relinquish their positions to nations investing in the future.

The fatal flaw of capitalist accumulation may well be inheritance. Neither the proclivities to strive and save, much less simple business acumen, run in family lines. But since most persons long to leave their riches to their progeny, we are perpetually burdened by large numbers of sloth-like individuals of subnormal ability who control, or at least benefit from, vast quantities of capital. Perhaps we need to complete the capitalist revolution. One of the main reasons for the brilliant success of corporate capitalism in the late nineteenth century was that it dismantled family organizations and began to operate by the more efficient principle of ability. Countries that adhered to the old-fashioned family model, such as Great Britain, failed to compete successfully with those developing managerial capitalism, such as Germany and the United States (Chandler 1990). As Michael Porter (1990:659) puts it, "Government can also influence the motivation of individuals through policies that provide citizens with *access to advancement based on merit*. The importance of this for economic upgrading is hard to overstate. Most entrepreneurs in America have not come from the upper strata of society." What in precapitalist societies is regarded as the virtue of looking out after one's own is in capitalism condemned as the vice of nepotism. If inheritance, which gives tremendous advantages to those who deserve nothing, comes to be considered as disgraceful as providing a half-wit son with a position of grave responsibility, we will have moved much closer to achieving the vaunted equality of opportunity that capitalism promises but never really provides.

But this remains an immoderate proposal that would be flatly rejected by virtually everyone who stands to inherit anything. George McGovern was crucified for suggesting as much in 1972. It also raises the thorny issue of control; who should gain access to a carefully husbanded supply of capital once the person who accumulated it perishes? Unfortunately, institutional investment agents, such as pension fund managers, have proven to be especially vulnerable to the lures of casino capitalism. The flaw may well be endemic to the system.

Conclusion

Considering the hostility of even many moderate environmentalists toward capitalism, the chance of effecting reconciliation between the environmental and the business communities appears to be slim. A vast gap of interest and expertise separates those with ecological and economic concerns, and until this gap begins to be bridged, progress will remain halting. With the notable exception of Alan Blinder, few liberal political economists have much of anything to say about the environment and, with the exception of the academic environmental economists (and the eco-marxists), few greens even want to understand economic processes. It seems quite possible that mutual hostility between the two camps will continue to thwart the development of a coordinated economic-environmental policy founded on a long-range vision that views ecological protection and economic prosperity as mutually reinforcing.

There is hopeful evidence, however, that a few environmentalists are tentatively beginning to endorse capitalism, just as a select group of capitalists are beginning to embrace environmentalism. It is the purpose of this work to encourage such a marriage. Yet a tremendous stumbling block remains, one that has received little attention thus far in this work. Many greens, relative moderates as well as radicals, remain convinced that contemporary capitalist success and general Western prosperity are largely the result of the exploitation of the world's poor countries. It is to this complicated issue that the final chapter turns.

6

Third World

Development and

Population

■ The Radical Position

Economic development and population growth in the poor areas of the earth are essential topics of environmental concern. Much of the so-called Third World suffers extraordinary—and rapidly accelerating—environmental degradation. The patterns of destruction experienced here are markedly distinct from those of the industrialized zone, calling for the development of a separate body of both social-environmental theories and economic-ecological programs. In the least developed countries, for example, toxic waste production is inconsequential (although such areas often serve as dumping grounds for the wealthy countries), but deforestation and desertification often proceed at devastating rates. Contrastingly, in the few Third World countries that are now industrializing successfully, toxic wastes are produced in massive quantities and are often disposed of far more casually than in the older industrial zone. In other words, both those countries that are successfully developing and those that are failing to develop often experience especially severe forms of environmental degradation.

A Third World?
One must take care to draw distinctions within this broad zone of global poverty. The environmental problems and prospects of Mexico, for example, are as different from those of Mali as they are from those of Germany. Still, terms such as "the Third World" or "the South" provide convenient labels for the earth's relatively poor countries. In this chapter

Third World will be employed to designate both the relatively nonindustrialized and the recently industrializing areas of the globe. The term admittedly obscures almost as much as it reveals, but such imprecision is necessary if we are to avoid using stiflingly cumbersome forms of expression.

With the notable exception of the primitivists, eco-radicals are as troubled by the human suffering of the Third World as they are by the ecological traumas the region faces. But their dual social and ecological concerns generate a paradox, for the economic development necessary to bring about broad-based prosperity has always been associated with rapidly increasing pollution and resource consumption. If the entire world were suddenly to catapult to an economic status congruent with that of Europe, Japan, and the United States, the environmental consequences would be appalling. Consequently, radical greens argue that the development necessary to relieve Third World poverty must be substantially different from the development that has characterized the rise of the First World.

Complexly intertwined with economic development is population growth. With few exceptions, Third World growth rates range from brisk (India and Guinea-Bissau at some 2 percent a year) to breathtaking (Kenya at 4.2 percent, Jordan at 3.9 percent a year), thus requiring substantial economic expansion merely for living standards to remain stationary. Yet through much of the region, economic growth has failed to match demographic expansion, resulting in deepening poverty and mounting environmental degradation. In impoverished and densely populated areas experiencing relentless population growth, such as Bangladesh and Egypt's Nile Valley, complete ecological and social breakdown in the near future seems distinctly possible.

Population growth, like environmental degradation, is itself closely tied to poverty. Mainstream environmentalists have long argued that overcrowding strains local resources, thereby impoverishing the local population; eco-marxists and other leftist scholars, on the other hand, typically respond that poverty leads to high fertility rates in the first place. Marxists also commonly contend that so-called overpopulation is usually just a symptom of the maldistribution of land and other resources. Traditional environmentalists thus typically advocate state-directed population control, while marxian-inspired writers argue that only genuine economic development founded on social equality can lead to demographic stabilization.

Considering the complexity of these issues, it is not surprising that the various eco-radical contingents should hold widely divergent opinions on Third World development and population growth. Let us now briefly explore some of these varying interpretations.

For and Against Development
The most extreme eco-radicals casually dismiss the very concept of development. If both social and ecological conditions were superior under regimes of subsistence agriculture—or better yet, hunting and gathering—then any step toward more complex economies will only result in the further degradation of both land and society. Moreover, since economic growth necessarily entails expanding levels of consumption, successful development will only further strain the already overburdened global ecosystem.

But even the most committed eco-radicals usually realize that a return to tribal subsistence is simply not possible, at least in the short run; population growth alone makes this quite impossible. Extremists of misanthropic bent thus openly hope that a massive increase in Third World mortality will diminish human pressures (see chapter one). More common is a vague despondency brought on by the realization of the fundamental incompatability of economic development and ecological survival. As Catton (1980:88) argues: "The new ecological paradigm was prerequisite to seeing that universal development was an unattainable goal. It was tragic that such a goal came to be universally sought. . . . Most of the poor nations of the world would never become rich."

But over the past ten years, most radical greens—to their credit—have come to regard the hope that nature will soon take its revenge against the teeming hordes of humanity as both bigoted and cruel, just as they have come to dismiss the despondent fear that genuine development is impossible as self-centered and ultimately self-defeating. The most widespread eco-radical position now is that Third World environments can only be preserved if poverty is alleviated through certain kinds of development initiatives. Third World peasants, they correctly argue, are forced to deforest and overgraze their landscapes precisely because of their poverty.

The challenge is thus to devise an environmentally benign method of improving living standards; to construct, in other words, a platform for eco-development. Such a program presents a radical departure from conventional development plans, which are roundly assailed as both ecologically destructive and socially regressive. Development in the old, liberal sense, argues eco-feminist Shiva (1989:80), should be seen as merely a "new project of Western patriarchy." Few deep ecologists would disgree.

The tenets of eco-development follow directly from the central propositions of mainstream (or deep ecological) radical environmentalism.

Development, accordingly, should be based on small-scale projects run by local communities and governed through participatory democracy. As Riddell (1981:159) argues, poor countries would be much better off if they elect simply to bypass industrialization. Instead, the rural production of craft goods should be stressed, and only technologies that fit local circumstances should be introduced. Production must be for subsistence rather than for exchange, and bioregions must strive for a self-sufficiency that will allow them to sever their ties with the global economy. These are, of course, the same prescriptions most eco-radicals offer the industrialized world. Their vision is thus one of global convergence, with the Third World developing in precisely the same manner that the First World de-develops.

Eco-development proponents generally insist that local societies also shun cultural connections with the larger world, especially those maintained through the electronic media. As Young (1990:197) explains: "[P]oor countries . . . would be well advised to refuse such inducements as commercial television, however much their elites may demand them. Like transistor radios, television sets will be used primarily to create new wants, only to be satisfied by the abandonment of subsistence production."

Eco-marxists and other radicals of more traditionally leftist leanings diverge profoundly from the standard eco-development course. They typically consider it, if nothing else, theoretically unsophisticated. W. M. Adams (1990), however, has recently attempted to unite the political-economic concerns of the left with an eco-development platform, hoping to devise in the process a truly radical developmental agenda. He is particularly anxious to discredit mere reformist sensibilities, such as he believes are evident in the program of sustainable development (see below), and to attack any strategies that aim to preserve nature only for the benefit of the elite. He also makes it absolutely clear that a properly constituted *leftist* version of green development will remain committed to anthropocentric goals: "ultimately . . . 'green development' has to be about political economy, about the distribution of power, and not about environmental quality" (1990:10). Adams diverges sharply from the marxian path, however, by insisting that development initiatives must come from below, focusing on the "aspirations of individual people" (p. 201). In the end he almost seems to be arguing that adequate development will spontaneously occur if only poor, rural peoples could be left to their own devices. "Green development" of such a variety thus returns, by a tortuous path, to a position remarkably close to that of the antidevelopmentalism of the primitivists.

Adam's proposals fit most closely into the humanistic anarchist tradition and are thus unlikely to receive support from either eco-marxists or mainstream eco-radicals. All three of these camps, however, unite in their acceptance of the neo-marxian theory that international capitalism is ultimately responsible for all forms of Third World poverty and degradation.

Notions of Dependency

Hopes for First and Third World convergence reflect a desire to redress past wrongs almost as much as a concern for environmental salvation. Most radical greens believe that the wealth of the industrialized countries essentially stems from their rapacious exploitation of the rest of the world. Through imperial conquest, Western nations (and Japan) arrogated the accumulated capital, the mineral and biological resources, and the human labor of the colonized zones. Such ill-gotten gains provided the essential fillip to Western industrialization while simultaneously ensuring the continuing impoverishment of the rest of the world. As Barry Commoner (1990:166) argues, global poverty is simply "the distant outcome of colonial exploitation. Colonialism has determined the distribution of both the world's wealth and its human population, accumulating most of the wealth north of the equator and most of its people below it."[1]

Virtually all eco-radicals concur, although few would frame the issue quite so simplistically. Moreover, most would further contend that the North's current prosperity and the South's current desperation remain directly and structurally linked. First World investment in poor countries, Porritt (1985:95) tells us, only results in further *under*development. Similarly, Rifkin (1989:232) argues that "as long as we in the United States continue to consume one-third of the world's resources annually, the Third World can never rise to even a semblance of a standard of living that can adequately support human life with dignity," while Bookchin (1989:11) informs us that the Sudanese could easily feed themselves if not forced by the United States—through its control of the World Bank and International Monetary Fund (IMF)—to grow cotton (see also Riddell 1981:79).

Our own industrial order thus becomes doubly objectionable, based on an ecologically unsustainable production system at home and on the unconscionable exploitation of lands and peoples abroad. Global development in the mode of the West is also revealed as impossible, since each developing region would have to exploit, and thus further impoverish, another even more marginal area. A few select countries might rise to the ranks of the exploitative powers, but a large and thoroughly impoverished periphery will necessarily remain.

The notion that Third World poverty derives largely from First World power is rooted in two schools of neo-marxist political economy: dependency theory (Frank 1969) and world systems theory (Wallerstein 1974). Although these two theoretical bodies are distinct, they share a number of basic attributes; moreover, the global theory propounded by most eco-radicals lacks the subtle distinctions that mark neo-marxian discourse. The following discussion, therefore, presents a brief and generalized overview of the dependency and world systems notions commonly employed by radical environmental writers.

According to neo-marxian models, the Third World's widespread poverty and lack of industry stem not from the failure of development to occur spontaneously, but rather from the more active processes by which the West undermined and forestalled indigenous economic growth. The imperial nations, in other words, underdeveloped their colonies, protectorates, spheres of influence, and nominally independent trading zones throughout the entire Third World. In the most blatant instances imperial agents demolished nascent industries in order to prevent them from competing with home producers. Elsewhere, colonialists undercut local industries and crafts more subtly through ruthless competition aided, when necessary, by gunships, armies, and not-so-subtle forms of diplomacy. Distant reaches of the globe were thus drawn into a world economy centered on the core nations of the West. The new global periphery was, in turn, relegated to the tasks of providing raw materials to, and purchasing manufactured goods from, the industrialized core.

According to some dependency models, the more fully a peripheral area was integrated into the global economy, the more subjugated, and hence impoverished, it would become. Although the configuration of core, semi-periphery, and periphery might fluctuate, the chance of genuine development occurring in the true periphery is virtually nil in the absence of complete global economic restructuring. Only those few areas that withstood imperialist pressures and maintained autonomy—such as Japan—have been able to embark on a path of genuine capitalist development (Moulder 1977).

Most eco-leftists further argue that the current structure of the world capitalist system ensures widespread hunger in the Third World. Since profit is the sole motive in a capitalist system, foodstuffs flow only to those who can afford them, not those who need them. Owing to the logic of capital, subsistence cultivation continually gives way to the fitfully profitable production of export crops, destined for the global core, that provides no sustenance for laboring populations. Similarly, eco-radicals argue that technical advances allowing increased harvests, particularly those associated with the so-called green revolution, result only in mounting landlessness, hence desperation, since only wealthier farmers can afford to use them in the first place.

Many proponents of dependency theory even argue that famines are largely a by-product of global capitalist economics, implying that hunger is virtually unknown in precapitalist and socialist societies. As capitalist social relations diffuse from the core and penetrate peripheral regions, previously existing social safety networks are steadily demolished. Such arguments can become quite intricate, if not disingenuous; Bradley and Carter (1989:113), for example, blame American capitalism for the Sahelian famine because American farmers sold wheat to the Soviet Union instead of donating it to the people of the Sahel. One might think that marxian communism would have to be considered *at least* as blameworthy in this instance, but the authors absolve it in one sentence, stating that only "individual Russian buyers" bear any responsibility. Again, the ills of capitalism are seen as systematic, while those of marxism are dismissed as accidental or idiosyncratic.

Dependency theory thrived in the 1970s when the locomotive of global economic expansion suddenly began to lose steam. Contrary to the expectations of liberal development theory, most Third World countries either stagnated or began to regress; the vaunted "take-off stage" of rapid development previously visualized now appeared to be little more than a mirage. Neo-marxian theories thus seemed to offer the most consistent explanations for the widespread failure of development to occur. For students and young faculty members radicalized in the late 1960s and early 1970s, the appeal of structural theories of global dependency was overwhelming.

Yet even as the gloomy doctrine of dependency began to sweep academia, rapid industrialization was occurring in several peripheral countries, most notably Brazil. Brazilian industrial growth was not marked by a decline in poverty for the country's most desperate regions and exploited communities, but it still defied the general expectations of dependency models. Even less explicable was the phenomenal expansion of South Korea and Taiwan. Here, at least, poverty was rapidly diminishing. By the mid-1980s a number of committed leftists began to reject the basic tenets of dependency theory. As Alain Lipietz (1987:4) writes of follow marxists who have taken the doctrine too closely to heart: "When researchers . . . adopt such attitudes they abdicate their intellectual responsibilities. Every aspect of a real social formation is seen as resulting from the evils of 'dependency.' Every concrete situation is forced into the Procrustian bed of a schema established by some Great Author of the past, while anything that won't fit is simply lopped off." Indeed, in recent years, according to a prominent theorist of an opposing camp, "criticizing dependency theory has become an academic industry of the worst sort" (Haggard 1990:19).

The intellectual poverty of the more simplistic notions of dependency will be examined in greater detail below. What is significant to note here is that few eco-radicals have followed the contemporary marxist and post-marxist debates. As a result, they cling to theoretical models that were popular years ago, unaware of how insecure their foundations have been revealed to be.

Demographic Growth

Perspectives on Third World population growth also divide environmentalists into several camps. To a large segment of the environmental movement, nothing is quite so worrisome (Ehrlich and Ehrlich 1990). According to this mainstream position, any advance in pollution control or habitat protection will inevitably be lost as a mounting human population consumes an ever greater proportion of the planet's resources. Many environmental writers, therefore, advocate strict demographic control, often based on naked coercion.

Radical greens of leftist bent, on the other hand, usually downplay Third World population growth. In fact, until fairly recently, "the anti-imperialist left dismissed this problem out of hand because it clung to an angelic conception of an ideal schema of reproduction: more mouths to feed meant more arms to feed them" (Lipietz 1987:146). Many eco-feminists apparently still accept this notion; a recent article in *Ms.* magazine, for example, tells us that "blaming environmental destruction on overpopulation obscures the main causes of the crisis" (May–June 1991, p. 83). Eco-marxists generally cannot afford such a cavalier attitude, but they often still attempt to deflect demographic concerns. First they argue that poverty itself causes unsustainably high rates of population growth in the first place, and second they claim that the affluence of the wealthy generates far more environmental damage than does the mere human weight of the poor. The average American, they correctly inform us, is responsible for generating several hundred times more industrial pollution than is the average citizen in Bolivia, Mali, or Nepal. Taking this reasoning to its extreme, overpopulation in the world's poorest societies becomes a trivial problem when compared with overconsumption in the world's wealthiest societies.

Despite such arguments, eco-leftists cannot simply ignore Third World population growth. Rather, most call for the global redistribution of resources in order to break the connection between poverty and fecundity: "the world population crisis . . . ought to be remedied by returning to the poor countries enough of the wealth taken from them to give their peoples both the reason and the resources voluntarily to limit their own fertility" (Commoner 1990:168). Commoner (1990:164), for one, sees a relatively simple and deterministic relationship between per capita wealth and fertility. Thus, he argues that if all countries could reach the level of prosperity enjoyed by Greece, the population crisis would be immediately resolved.

One of the main reasons eco-marxists are not terribly concerned about population is that they typically regard habitat preservation as a low priority. In many Third World countries, and certainly in most of sub-Saharan Africa, human numbers can indeed increase for some time without directly threatening the resource base necessary for economic development. This by no means implies, however, that increasing demographic pressure will not detract from the reproduction of other species. Indeed, many nonmarxian environmentalists worry about high fertility rates in the Third World precisely for this reason.

Combining the Affluence and the Overpopulation Arguments
Recently, Paul and Ann Ehrlich (1990) have attempted to accommodate the leftist "poverty thesis" into their own neo-Malthusian scenario of the soon-to-explode "population bomb." According to the Ehrlichs, population pressure is already at unsustainable levels virtually the world over. Continued demographic expansion, they argue, is the global environment's greatest threat. In a unique departure, however, they now contend that both wealthy and impoverished societies contribute to the problem in roughly equal measures. Environmental impact, they claim, is equal to a given country's population multiplied by its index of affluence and by its level of technology (in abbreviated form, $I = PAT$). Each (average) American, given his or her standard of living and access to technology, is thus ecologically equivalent to 280 citizens of Rwanda. Yet both countries are described as growing more grossly overpopulated with every passing day—Rwanda because its population (P) is expanding, and the United States largely because of its technological (T) advances. The Ehrlichs thus indict the overly affluent First World for maintaining (rough) population stability when in fact it should begin a rapid demographic descent in order to make up for its abnormally high levels of "A" and "T."

Yet a few wealthy countries have already started to experience population decline. Even American fertility rates are below replacement level, and moderate demographic expansion is occurring here only because of immigration. To most observers this is irrefutable evidence that the First World is not experiencing a population explosion. But by equating human numbers with affluence and technology, the Ehrlichs can argue the contrary. They thus regard the birth of every American baby as a rather tragic event (1990:10), and they consider couples who choose to have more than two children as abhorrently immoral. To counter the unsustainably high fertility rate of 1.9 children per American woman, the Ehrlichs (1990: 189) advise severe indoctrination: "happy, successful families in classroom stories and films should never be shown with more than two children." The resulting infringement of freedom is held to be a minor inconvenience; the very survival of the planet, after all, is at stake.

■ The Necessity—and Possibility—
of Economic Development

As most greens realize, economic development of some sort is necessary for both social and environmental reasons. Political ecologists have demonstrated over the past decade that the poorest peoples in the poorest countries are increasingly forced to eke out their livelihoods from ever more marginal lands, resulting in deforestation, soil erosion, desertification, and other forms of habitat destruction. The lack of economic alternatives also leads Third World political elites to encourage the unsustainable harvests of living resources. Only though genuine development can such pressures be reduced.

The relationship between population growth and economic development is more complex. Despite the assertions of a few leftists, population stability does not always neatly correlate with economic development as defined by such crude but irreplaceable measures as per capita GNP. The wealthy oil exporters of the Persian Gulf, for example, are burdened by exploding populations (Qatar is expanding at a remarkable annual rate of 4.1 percent), whereas impoverished Sri Lanka has experienced a remarkable fertility decline (Sri Lanka is now growing at some 1.3 percent a year). But a less precise correlation between national prosperity and demographic stabilization does still obtain, bolstering the notion that economic development is necessary for environmental protection.

The essential challenge is thus one of devising an ecologically benign developmental pathway that will lead to population stability. Let us begin by examining the deep ecologist's proposed panacea, usually labeled eco-development.

From Eco-Development to Sustainable Development
Eco-development, as outlined above, rests on the same basic propositions of radical environmentalism that have already been rejected in
this work. No program committed to small-scale technology and economic autarky can ever foster genuine development. Moreover, just
because a project is small in scale by no means implies that it will
be environmentally benign. A prime example would be China's 1959
"Great Leap Forward," a program predicated on rural iron production.
Although eco-radicals such as John Young (1990:183) automatically endorse this scheme because of its small-scale nature, careful scholars have
shown that it was nothing less than an environmental disaster (Smil
1984).

Seeing the pitfalls of eco-development, a group of nonradical scholars
has recently forwarded the concept of sustainable development as a
means of addressing Third World poverty while ensuring environmental
protection. Not surprisingly, sustainable development has been excori-
ated by right-wing thinkers for being hopelessly radical (Anderson and
Leal 1991) and by leftist thinkers for being hopelessly reformist, hence
conducive to continued exploitation (W. Adams 1990.)[2] In reality, it is
neither. Although sustainable development's initial premise, that "the
natural capital stock should not decrease over time" (Pearce et al. 1990:
1), appears congruent with the radical opposition to the extraction of
nonrenewable resources, the concept's key terms are defined broadly
enough to allow both resource use and economic growth (see Pearce et al.
1990:10). It turns out that the main premise of sustainable development
is that economic growth must never undercut the productivity of natural
ecosystems. In fact, some sustainable development enthusiasts advocate
relatively fast economic growth in the Third World, provided that it is
accompanied by a "significant and rapid reduction in the energy and raw
material content of every unit of production" (MacNeill et al. 1991:24).
This is a platform on which Promethean environmentalists readily
stand.

Nor is Promethean environmentalism necessarily opposed to all eco-
development programs. Carefully devised, environmentally aware,
small-scale development strategies, whether labeled "eco" or "sustainable," do form an essential interim approach for reducing rural poverty.
Small-scale development projects, especially those that work through
the ecological knowledge of indigenous peoples, can both safeguard local
landscapes and significantly improve living conditions. In most areas of
the Third World only ecologically oriented agrarian development programs can offer *immediate* hope for impoverished cultivators, and thus

begin to sever the links between deepening immiseration and accelerating degradation.

Eco-development programs should thus be embraced, but only in conjunction with policies ultimately aimed at industrialization. Unfortunately, even sustainable development advocates usually focus exclusively on rural, agrarian issues, ignoring the larger urban-industrial nexus. But ruling out industrialization would only condemn the Third World to perpetual poverty and subservience to the now industrialized countries. Although radical greens would counter that global equity should be achieved by deindustrializing the First World, such a fantasy will not likely ever be realized. Third World poverty, on the other hand, is a grinding reality.

Even from a strictly environmental perspective, any development program that denies the need for industrialization and urban development will ultimately prove self-defeating. Mere eco-development would, for example, preclude the Third World from enjoying the many environmental benefits provided by urbanization (see chapter three). As ecologist John Terborgh (1989) argues, urban growth in Latin America offers the only hope for habitat preservation. If rural populations continue to expand, the large blocks of wild vegetation necessary to ensure biotic diversity will inevitably diminish. Third World urbanization also promotes a shift away from utilitarian environmental attitudes. Popular ecological movements in the Third World are far more widespread in cities than in the countryside, and they usually rely substantially on middle-class support.[3] Finally, but most importantly, it is essential to recognize that even the most successful rural development projects are not generally accompanied by plunging fertility rates; urbanization, on the other hand, usually is.

If Third World urban-industrial development offers long-term environmental advantages, it also clearly entails some spectacular short-term environmental costs. Early industrialization is accompanied by horrific pollution, forming an ecological assault that eco-radicals feel the earth could not tolerate. Although such fears are overblown, more benign forms of industrial production must be applied in the future. Most importantly, the sophisticated, environment-sparing forms of technology outlined in chapter four must be developed and transferred to the Third World as rapidly as possible. Ultimately, as Reddy and Goldemberg (1990:115) argue, Third World countries should strive to engage in "technological leapfrogging" by adopting "more energy-efficient technologies even before they have been widely adopted in industrialized countries." Whether this could occur quickly enough to spare the global ecosystem is an open question; both rapid innovation and substantial state subsidies will probably be necessary. Technology transfer will also require major assistance from governments in the First World—a region increasingly preoccupied with its own economic and environmental problems, and remarkably stingy in sharing its industrial secrets (see Porter and Brown 1991).

Another legitimate concern is that heavily polluting industries will increasingly flee countries with stiff regulations to settle in poor countries with lax environmental standards. Indeed, some states (most notably Romania before its anticommunist revolution [Leonard 1988]) have sought to attract filthy industries, apparently unconcerned with their effects on the local environment. Although a few economists have argued that this form of pollution exporting may help poor nations industrialize, environmentalists recognize that it would also undermine regulatory attempts to improve environmental health in the First World. Overall, however, the "pollution haven" industrial transfer process has proved but a minor force in global economic restructuring (Leonard 1988). More worrisome is the export of toxic wastes and other hazardous materials from the First to the Third Worlds; on this score the actions of multinational corporations and First World governments have often been scandalous (Scherr 1987).

Urban Development Reconsidered

Critics of Third World urbanism and industrialization also point to the appalling environmental conditions of such overcrowded, industrial cities as São Paulo, Bangkok, and Mexico City. Urban development has already progressed too far they argue, spawning virtually uninhabitable megalopolises. But while the environments of such cities are wretchedly degraded, the question remains as to whether this stems from urbanism per se or from the poverty that accompanies low levels of economic development. London in the nineteenth century was a foul place, but as it became more prosperous, while still expanding, it grew steadily cleaner. Modern Tokyo is as large and as industrial a city as any, and if its environment is not the world's most pleasant it is certainly tolerable— and far less polluted than it was twenty-five years ago.

Several Third World metropolises admittedly face special environmental challenges that may not be solved by economic development alone. Mexico City, for example, confronts severe difficulties merely by virtue of its location in a high-altitude basin. Human numbers in the Valley of Mexico may already exceed the restorative capacity of the local environment. The Mexican government is thus wisely attempting to shunt further urban growth to secondary cities (Leonard 1988:143). But the pull of the capital city remains powerful, and, as government planners recognize, state functions themselves must be transferred to other cities if urban decentralization is to occur.

The hyper-urbanization of cities such as Mexico points to an important aspect of Third World urban development usually neglected in the environmental literature. In many instances, the recent explosive growth of urban areas reflects not genuine economic development but rather the developmental distortions generated by a combination of economic stagnation and over-bureaucratization. In the absence of true development, the central government often becomes the single pole around which urbanization occurs. Most Third World countries are characterized by urban primacy; they are dominated, in other words, by a single, massive city. In a genuinely developing society, in contrast, separate urban areas are economically integrated into systems of cities that generally develop in concert.

Urban primacy is for good reason regarded as unhealthy, and many Third World countries are wisely encouraging the growth of secondary and tertiary cities. Urban decentralization of this sort is environmentally and socially desirable, but it is notoriously difficult to implement through fiat or even careful planning. Achieving economic development across a broad range of industries is ultimately necessary for the emergence of a healthy urban sector.

One must also recognize that the hellish conditions of many Third World cities reflect political failure as much as economic stagnation. In most of these cities elites have struggled to maintain the pleasant, low-density environments of their own posh neighborhoods, and in so doing they have forced the poor into horrendously cramped quarters. In seeking to staunch rural-to-urban migration, the wealthy have often prevented the disadvantaged from acquiring property rights, thereby discouraging private investment in impoverished neighborhoods. In seeking to maintain "visual aesthetics," the powerful have on countless occasions ordered the demolition of shanty settlements, thereby condemning thousands at a time to privation. And in shortchanging public investments, often in favor of vanity projects, political elites have failed to provide basic infrastructural services. Even in the world's poorest countries, urban sewer and water systems are within the scope of national economies—if only power holders (and international lending authorities) would recognize their importance (on political neglect in general, see *The Economist*, "Africa's Cities: Lower Standards, Higher Welfare," September 15, 1990, pp. 25–28).

Finally, one must realize that the problems faced by existing Third World cities could never be solved by rural eco-development projects. Short of a Khmer Rouge-style global deurbanization program, urban problems—requiring urban solutions—will persist.

Promethean environmentalism accepts the desirability—indeed, the necessity—of urbanization throughout the world. It seeks strenuously, however, to enhance *genuine* urban-social development. Essential to this project is the social mainstreaming of marginalized urban settlers. As Hardoy and Satterthwaite (1985:206) write: "The developing city is increasingly becoming an illegal city built by the efforts of people, many of whom were born, will live, work and die without a record." Guaranteeing basic human rights, in the end, will prove essential for both economic development and environmental protection.

Healthy urbanization thus demands both the provision of human rights as well as industrial development. Most eco-radicals, however, disparage industry, and even those few who recognize its value usually counter that peripheral industrialization is impossible in a capitalist world system. The present task is thus to reexamine the theory of dependency that influences virtually all forms of eco-radicalism.

Development and Dependency

As noted above, the dramatic success of South Korea and Taiwan undermines the main supports of the less sophisticated versions of dependency theory (Corbridge 1986). While some authors would dismiss the rise of these nations due to the favors they received from a United States anxious to stem the tide of communism, the equally remarkable success of Hong Kong and Singapore, as well as nascent industrial growth of Thailand, Malaysia, and Indonesia, indicates that the phenomena of Asian development is indigenous. It is true, however, that Taiwan and South Korea were, until recently, heavily dependent on the United States. But it is also clear that they were able to manage their dependency and eventually transcend it (Stallings 1990). As Ellison and Gereffi (1990:398) conclude, after rigorously surveying the economic evolution of a wide variety of developing countries, "[L]inkages to the world economy thus can have positive as well as negative consequences for national development, depending on how and when they are established, and whether they are congruent with key elements in the domestic environment."[4]

The crude version of dependency theory popular in eco-radical circles also fails a significant test when one examines countries that were never subdued by the imperial powers. Afganistan, Nepal, and Ethiopia were lightly touched by European colonialism and they attracted scant attention from Western merchants. Dependency models would lead one to suspect that by maintaining autonomy such areas would have been able to lay the foundations for genuine development and that today they should be among the more prosperous of Third World countries. In actuality they are among the most impoverished. Such poverty cannot be attributed to their remote and rugged locations; all were sites of old civilizations that were once as prosperous as, if not more prosperous than, surrounding regions. Similarly, dependency models would lead one to suspect that countries such as Burma (Myanmar), Guinea, and Madagascar that have more recently isolated themselves from the world economy should have fared better than those countries that opened themselves to the rapacious firms of the imperialist powers, such as Thailand and the Ivory Coast. But again this notion is belied by recent history. Out of desperation the world's economic hermits are one by one asking for the return of foreign capital.

The economic records of Third World communist countries are no better. Not only have they failed to develop, but even the marxian excuse that they have at least eliminated hunger is overstated. As Hamerow (1990:322) shows, "more than a decade after the victory of the revolution in Vietnam, the government acknowledged that serious malnutrition threatened the nation because of a drop in food production and a rise in the birthrate." Eco-marxism cannot even begin to explain why the Vietnamese government now pleads for a reentry of foreign capital. A dozen years ago enthusiasts could still write about optimal industrial growth in socialist Third World cities (Forbes and Thrift 1981:14–15), but such a notion now appears quite ludicrous.

Yet the enthusiasm of Western radicals for marxist states like China can still verge on the pathetic. William Bunge (1989:356), for example, argues that "We could create societies where children bloom like little flowers, rather than die like flies—as the Chinese are demonstrating." If one wants to learn what is really happening to Chinese "little flowers," I would suggest reading Steven Mosher's (1983) Broken Earth. I would also ask Bunge where he thinks one might find the missing 600,000 Chinese girls born last year; many demographers, after all, strongly suspect infanticide (New York Times, "A Mystery of China's Census: Where Have the Girls Gone," June 17, 1991, p. 1).

Imperialism and European Prosperity
Even deeper flaws are apparent in the larger argument, propounded by Barry Commoner, that the success of the West is merely a result of its exploitation of the Third World. Although it is true that Western imperi-

alism inflicted drastic damage on the economies and environments of all colonized areas, and that exploitation of the periphery gave a profound boost to early industrialization in the imperial heartland, this is by no means the entire story. Ultimately, it is most reasonable to adopt a balanced view that sees First World prosperity and Third World poverty as related both to internal socioeconomic dynamics and to the history of external exploitation and oppression. In practice, however, weighing the relative importance of these two factors is almost impossible.

Determining the role of plunder and colonialism in generating Western prosperity is complicated by the intricate history of Europe's imperial grasp. In the initial phase of Iberian expansion, colonial gold and silver brought riches to the Spaniards, but they also undermined any previously existing developmental impetus (Ringrose 1989). The Portuguese, for their part, found that the costs of policing their far flung empire consumed most of their trade gains. Imperial wealth did invigorate European commerce, however, just as it enriched trading, banking, and manufacturing regions, such as the Spanish Netherlands. Yet the general upswing of the European economy began well before the voyages of exploration and conquest returned home with ill-gotten gold. In the seventeenth century Dutch mercantile imperialism, based on a tight-fisted control of Eurasian maritime trade routes, certainly yielded what Simon Schama (1988) calls an "embarrassment of riches." On the other hand, it by no means allowed the Dutch to "hatch the industrial chicks" and thus begin the cycle of self-perpetuating economic growth (Schama 1988:6). Moreover, the Dutch "mother trade" was intra-European, based on control of the Baltic. As the Netherlands' European position declined in the eighteenth century, so did Dutch prosperity—despite its vigorously expanding colonial holdings in what is now Indonesia.

Throughout the sixteenth, seventeenth, and early eighteenth centuries, Europeans dominated only isolated ports through most of Africa and Asia. Land-based empires were still untenable in the former continent due to disease and in the latter due to the military power of the indigenous states. Only with Great Britain's conquest of Bengal in 1757 did European powers begin to build continental empires in the Eastern Hemisphere. The treasure hoard of Bengal was a tremendous windfall for Great Britain, and it is an intriguing coincidence that England's industrial revolution began at roughly the same time as its piecemeal conquest of South Asia commenced. It is also essential to note that imperial sugar plantations, usually based on slavery, provided the cheap calories needed to sustain the lives of the heavily exploited workers in England's early textile mills (Mintz 1985). But few scholars of the industrial revolution see the plundered wealth of India, much less the flow of sugar, as early industrialism's *prime* mover.

The nearly complete division of the earth into colonial territories and spheres of influence did not occur until the late nineteenth century, well after the leading imperial powers were thoroughly industrialized. Marxist theory has long accorded this overseas thrust primarily to capitalism's incessant overproduction and corresponding need for captive markets, and secondarily to its expanding requirements for raw materials. Michael Doyle (1986), however, in his careful study of the history of imperialism, shows that while economic factors cannot be discounted, military, political, social, and cultural forces played equally important roles. The scramble for Africa, in particular, owed as much to intra-European national pride as to any real or perceived economic advantages to be gained. As John Hall (1985:224) concludes: "It is not possible to argue that colonies were required as places in which to invest capital for the simple reason that capital outflows from France went overwhelmingly to Russia, where good profits could be made, rather than to her own colonies. . . . Nor did French industry send massive amounts of exports to its colonies, most of which, being poor, were not able to afford them."

Penny-pinching mercantile capitalists usually preferred to obtain markets and materials through trade, unaccompanied by imperial expansion. Great Britain commercially penetrated independent Latin American countries more thoroughly than many of its own African colonies, a process that was, admittedly, highly destructive to local producers. By the late nineteenth century, after its manufacturing advantages had declined, British interest in Latin America began to shift toward infrastructural investment. Argentine railways, in particular, seemed an attractive opportunity. By investing capital, dependency theorists argue, European financiers distorted and undermined Argentina's economy for the benefit of the imperial states. It is, however, difficult to place all the blame for Argentina's current economic woes on such investments; after all, another favored target for British capital exporters was the United States.

The overall impact of imperial expansion, both formal and informal, on the European home economies is thus almost impossible to assess. Great Britain surely profited greatly from its trading and plundering activities, and its control of the Indian army allowed it to maintain a global empire at an astoundingly low cost (Washbrook 1990). But the long-term consequences of its early successes are by some accounts surprisingly negative. Michael Porter (1990), for example, convincingly argues that Great Britain's possession of a large and captive colonial market lulled its entrepreneurs into a deadly complacency. In contrast, American manufacturers, selling initially within a domestic market, made steady gains in competitive advantage. And, despite high rates of return, British investment in Latin America may have ultimately undermined the British economy. Starved for the capital that increasingly flowed overseas, many British firms proved unable to keep pace with foreign competitors (Rosecrance 1990:56).

An interesting perspective on the role of colonialism in European economic history may be gained by comparing Great Britain and Germany. As a quintessential maritime nation and the first industrializer, Great Britain easily maintained its standing as the foremost imperial power. Germany, by contrast, missed the early waves of imperial aggrandizement, and while it was a prime instigator of the late nineteenth-century global land rush, its spoils were relatively insignificant. Germany brutally exploited its poor territories, but for meager rewards. Then, after the First World War, the Germans were divested of all overseas possessions. Yet such imperial failures did not undermine the German economy. By the time Africa was partitioned, Germany was already ahead of Great Britain in key industrial developments, and despite its devastating military defeats it has never had to look back. Indeed, Germany's lack of colonies forced it to devise technological substitutes for raw materials (Hugill 1988); chemistry, it has been said, was Germany's substitute for empire. Adversity, as always, has its advantages.

Wellsprings of Prosperity

If the West did not grow rich solely by plundering the rest of the world, what then were the secrets of its success? This is, of course, a wearingly complex and controversial topic, and I can only offer here a few of the more cogent explanations offered in recent years. Rosenberg and Birdzell (1986) stress innovations in trade, technology, and social organization, arguing that the wide diffusion of power and the lack of strict religious restrictions offered vast room for economic and technical experimentation. Michael Mann (1986:412) locates Europe's rise in the Middle Ages and links it both to "the multiplicity of power networks and the absence of monopolistic controls" and to the "extensive networks and pacification provided by Christendom." Like Rosenberg and Birdzell, Mann stresses the diversity that characterized all spheres of European life. Similarly, Hall (1985) emphasizes the autonomy of power holders in traditional European civilization, and concludes that it was the miraculous combination of "commerce and liberty" (1985:249) that unleased the forces of development. Finally, Joel Mokyr (1990) convincingly urges that we not downplay European technological innovations.

Although these theories deserve careful consideration, they all err in overstressing European exceptionalism; in today's world it is the rise of the East that demands explanation. The most impressive attempt to understand economic success in a non-Eurocentric manner is E. L. Jones's (1988) *Growth Recurring*. According to Jones, all societies have the potential for realizing intensive economic growth (characterized by increasing per capita wealth), but that in most instances rent seeking elites have skimmed off so much produce that only extensive growth (characterized by expanding populations or territories but stagnant income levels) is possible. In the few historical societies in which governments were neither too strong nor too weak, thus allowing economic forces room to operate (notably Sung China, Tokugawa Japan, and early modern Europe), genuine development occurred. According to this line of reasoning, the West triumphed largely because conditions conducive to intensive growth persisted over a large enough area and for a long enough time period that it could simply outdistance the rest of the world.

Just as the West did not rise merely by piracy, neither does its current prosperity derive fundamentally from its exploitation of the lands and peoples of poorer nations. Global trade and resource flows now largely connect the various wealthy zones with each other. Despite the tenets of several new international division of labor theories, the direct foreign investments of U.S. firms have been increasingly directed to Europe (Schoenberger 1990:380), where labor costs are high and safety and environmental standards stringent, rather than to the Third World, where labor is cheap and standards lax. The contemporary complaint of the world's poorest countries is not so much that they are being exploited but rather that they are being ignored. Whereas fifteen years ago cutting-edge marxists decried the dependency-generating flow of capital to the Third World, today they bemoan the fact so little development-generating capital is being transferred (Thrift 1989:31).

Contemporary Exploitation: Paradoxical and Marginal
This is not to imply, however, that the First World does not, in many real respects, exploit the Third World. But much of the exploitation that occurs has decidedly ambiguous consequences. Affluent American consumers, for example, benefit when they purchase clothing sewn by Bengali workers who are paid a bare survival wage, a relation that can hardly be considered nonexploitative. But if we were to cease importing garments from Bangladesh, Bengali workers—and indeed the entire Bangladeshi economy—would suffer mightily. Cut off from export markets, Bangladesh would relinquish the small chance it now has for successful

development, while its industrial workers would be forced into even more precarious positions as landless rural laborers. Many would probably join the exodus to Assam or the Chittagong Hills, thus displacing indigenous cultivators and destroying precious wildlife habitat.

The First World also benefits from the destruction of nature in the Third World. The tropical rainforests of Southeast Asia, for example, are being stripped so that wealthy consumers (primarily Japanese) can purchase cheap wood products (Laarman 1988). But such destruction is inconsequential to the national economies of the developed countries. In the timber-exporting countries, concession-holding elites may grow stupendously wealthy, but very little genuine development results from their activities. Ultimately, the draining of biological resources from the Third to the First Worlds does far more damage to the regions of extraction than it confers in advantages to the regions of consumption.

The wealthy countries also import large quantities of agricultural produce from the Third World, although less in relative terms than they once did. This is an especially paradoxical issue. Eco-development enthusiasts routinely denounce cash-crop exporting; by growing for the global market rather than for local subsistence, they argue, peasant communities sacrifice their ability to provide themselves with sufficient foodstuffs. The result, in the worst cases, is widespread malnutrition if not outright starvation.

Although the linkage between cash-crop production and malnutrition is well established, it remains true that national economic development requires the generation of foreign exchange. Otherwise, capital goods and, in most cases, adequate energy, could not be imported. Many Third World countries have had little option but to export agricultural produce. In fact, when wealthy nations cease to import farm commodities, the resulting economic dislocations are often devastating. America's boycott of Nicaraguan coffee during the days of the Sandinista regime was roundly denounced by American leftists for precisely this reason. Similarly, the proximate cause of the utter immiseration of peasants and landless workers in sugar exporting zones, particularly the island of Negros in the Philippines, is America's declining import quotas—the result of a policy aimed at protecting cane, beet, and corn growers in the United States. Certainly elite Philippine landlords are equally to blame, as many of them forbid their former workers from growing subsistence crops on the now idle lands. But if a return to subsistence cultivation would lead in the short term to fuller bellies, it would still never generate genuine development.

If global economic development were the sole concern, one would have to argue that all import quotas and tariffs protecting agricultural and natural resource markets in the industrialized nations should be abolished. Ecological requirements, however, call for different policies. Buoyant export markets in agricultural and forest products lead inevitably to the conversion of natural habitats to biologically impoverished farmlands, tree plantations, or, in the worst cases, cut-over wastelands. The degree of degradation varies with different crops; sugar cane is a near total disaster, but shaded coffee plantations can actually support a healthy diversity of avifauna (Terborgh 1989:144). Yet it remains clear that an urban-industrial model presents a developmental path environmentally preferable to one based on the export of agricultural produce. Free trade in agricultural commodities proves to be an ecologically risky proposal indeed.

The Legacy of Imperial Exploitation

The preceding pages argue that (1) genuine development can and does occur in the periphery of the modern capitalist world system; (2) the wealth of the First World does not rest *fundamentally* on its exploitation of the Third World; and (3) exploitation itself often grows fuzzy when one considers its larger economic ramifications. But it cannot be denied that the poverty of the Third World can be traced *in part* to the damage inflicted by colonization. Just because a few former colonies have now achieved prosperity by no means absolves imperialism from its sins.

Dependency theorists are, for example, correct in arguing that the colonial powers often purposefully deindustrialized the areas they subdued. The British demolished early textile factories in Egypt (Bernal 1987:246–50), while all Western powers commonly disabled the more advanced local economic endeavors. Factory owners at home greatly feared colonial competitors who could profit from their access to cheap labor and raw materials. Owing to their influence over the metropolitan governments, industrialists were able to manipulate colonial policy to their own advantage. Here, it might seem, capitalism did indeed stifle Third World development.

The problem with such reasoning, however, is that it again fails to differentiate the logic of capitalism in general from that of the individual firm. English textile manufacturers acted in self-interest in demanding the destruction of colonial competitors, and the British government certainly acted in what it perceived as its own self-interest in complying. But such actions were entirely contrary to the spirit of capitalism, which never respects national boundaries or cultural distinctions. According to capitalism's advocates, competition is *always* a bracingly positive influence, even if it is *always* feared by vulnerable firms. But pure capitalism never exists; any given economy is always, to a greater or lesser extent, subordinate to national political interests. In most colonial settings individual capitalists and imperial agents successfully conspired against capitalism as a system.

The former Japanese colonies of Korea and Taiwan form an instructive counterexample to the process of colonial deindustrialization. Unlike the Western powers, the Japanese decided that it was in their best interests to industrialize several of their possessions (Myers and Peattie 1984). This was by no means done to improve local living standards; Japanese imperialism was as brutal as any. But the policy proved beneficial in the long run, and it is not coincidental that these two countries have been the most successful Third World industrializers. Still, it is far from the whole story. Japanese industrial investments in Korea were highly concentrated in the north, an area that has lagged woefully behind the formerly agricultural south.

In the final tabulation imperialism probably caused more damage to the colonized zone than it conferred in benefits to the colonizing powers. Although the legacy of imperial conquest and the realities of contemporary First World global economic domination do not explain the poverty of the Third World, they do form significant components of the story. The wealthy nations, particularly those that were the most brutal imperialists, thus owe a great debt of expiation to their former colonies. This could be partially accomplished by releasing poor countries from their debt burdens, a surprisingly inexpensive proposal (Krugman 1990:149–51) that could actually benefit the First World by increasing the global demand for capital goods and other industrial products. (As one hopeful sign, a debt-relief plan for Mexico has already given the country "the breathing space it needed to tempt back flight capital and boost GDP growth" [*The Economist*, April 27, 1991, p. 82].) The industrialized countries should also greatly accelerate the transfer of technologies, particularly those with environmental benefits, to the underdeveloped world. Importantly, this is exactly what many Third World countries are demanding before they agree to any sort of global environmental bargain aimed at protecting the atmosphere (Porter and Brown 1991). Shamefully, the United States, in particular, has steadfastly opposed all such proposals.

It is also reasonable to insist that wealthy nations are particularly obligated to assist their own former colonies. The United States thus owes its greatest debt to the Philippines, a nation that we victimized in myriad ways (Schirmer and Shalom 1987)—even if we were relatively benign colonialists, at least when compared to the French and the British. Yet the United States at present seems perfectly willing to abandon the Philippines to a descent into Bangladesh-style immiseration.

If we can no longer accept the notion that poverty in the Third World is strictly a result of imperial exploitation, then we are challenged to devise a more encompassing theory. Only by understanding global poverty can we begin to devise workable developmental strategies, which, in turn, are necessary if we are to ensure planetary survival.

Capitalism and Mercantilism in the Third World

Most eco-radicals agree that peripheral capitalism in the Third World, subservient to the requirements of the metropolitan states, never results in genuine development. One of the many problems with this line of reasoning is that it incorrectly assumes that most Third World economies can be unproblematically characterized as capitalist. On the surface most are indeed capitalistic, founded as they are on the private ownership of resources and on markets in land, labor, and capital. But the forms of capitalism present in the underdeveloped world are fundamentally different from those that have launched the wealthy nations on a path of sustained growth, a difference that stems primarily from indigenous, not exogenous, forces. It is best to relinquish our view of a monolithic capitalism and instead inquire how specific political-economic circumstances within developing (or, as is unfortunately sometimes the case, de-developing) countries have undercut economic progress. To do so is not to place all blame on internal circumstances; as the preceding discussion makes clear, imperialism must never be absolved. But it is also counterproductive to use imperialism as a universal scapegoat and thus deflect attention from political and economic failings located within Third World societies.

One of the more powerful challenges to the notion that the Third World is fundamentally capitalistic may be found in the work of the Peruvian economist Hernando de Soto (1989), who typifies the so-called neo-liberal school of Latin American political economy. De Soto argues that Peru, like most poor countries, is more accurately characterized as mercantilist than capitalist; in other words, its political economy is strongly reminiscent of Europe's prior to the capitalist revolution of the late eighteenth and nineteenth centuries. A mercantilist polity is characterized by market relations to be sure, but in a manner so distorted and stifled by bureaucratic regulation that a true market economy does not exist. Under such conditions, state bureaucracies seek primarily to perpetuate themselves, regardless of the economic harm they may cause in doing so, while an established merchant and industrial elite manipulates state agencies in order to shelter itself from competition. Both the bureaucracy and the established elite, de Soto argues, consistently deny commercial opportunities to the poor, whom they view as potential threats to the established order. The result is a kind of perverted socialism for the rich that leads invariably to economic stagnation.

One result of mercantile distortion is the emergence of a large informal (in other words, illegal) economic sector. Poor entrepreneurs are forced to operate outside of the law, since the time and money required to gain official sanction are well beyond their means. But as informals they lack property rights, preventing them from expanding and gaining economies of scale. De Soto's proposed solution is simultaneously to "formalize the informals," thus providing them the legitimate status needed if they are to compete with recognized operations, and to "de-formalize the formal sector," thus lifting from it the burden of capricious regulation that renders it uncompetitive. For this to occur, de Soto argues, both extensive judicial reform and a thoroughgoing democratization of political power are necessary.

De Soto presents his work as a challenge to both the Peruvian left and right. According to him, both camps—although they are loath to admit it—share a fundamental set of political-economic beliefs. As he argues (1989:xxvi), "The policies of both [the left and the right], at least in Peru, reinforce the mercantilist order. The right pursues mercantilism in order to serve particular business groups. The left does so under the illusion that it is benefiting the needy." De Soto also confronts the economic establishments of the wealthy countries which, he argues, have consistently stressed economic liberalization at the macro level only, ignoring the deeply rooted mercantile structures that consistently thwart development.

By championing the poor not as state wards but as entrepreneurs, yet at the same time viewing them not merely as atomized individuals but also as community organizers, de Soto provides a challenging and innovative model of development. But the model is not without serious flaws. Leftists dismiss it for ignoring class, and they are indeed correct in noting that the structural barriers to Latin America's poor are much more profound than those merely of bureaucratic regulation. And as Robert Klitgaard (1991) demonstrates, the weakly developed information infrastructures of most Third World countries preclude the development of efficiently operating markets; concerted state-level intervention, he argues, is necessary *before* adequate market mechanisms can be established. We might also object to de Soto's historical analogy. Many mercantile regimes of precapitalist Europe actually proved quite successful in nurturing productive enterprises, while both South Korea and Taiwan may be referred to as mercantilist polities without doing injustice to the term.

But regardless of such difficulties, de Soto's contention that we are ill served by viewing Peru as a purely capitalist country still stands. Moreover, the same argument may be made for most at the other nonsocialist Third World countries. The Philippines, in particular, provides an instructive example.

In the 1960s and 1970s Ferdinand Marcos attempted to create in the Philippines an authoritarian, technocratic, export-led, agro-industrial economic system, much on the model of South Korea and Taiwan (Hawes 1987). Marcos first had to contend with a wide variety of bourgeois classes (landed elites, import-substitution industrialists, exporters), all of which had different interests and pulled the state in different directions. After declaring martial law in 1972, Marcos claimed the power needed to reconstruct the economy as he saw fit, with the stated aim of bolstering exports, particularly of agricultural products and raw materials. To accomplish this, he transferred economic power from private to state hands, strengthening tremendously the sector of state capitalism. Eventually he established a system known locally as crony-capitalism, which quickly proved to be a parody of the genuine thing. Marcos would simply destroy firms owned by individuals perceived as enemies while subsidizing companies owned by supporters. That such an arrangement favors political over economic endeavors, undermines competition, and thwarts investment should be obvious. It also destroys any sense of national purpose. As Hawes (1987:142) writes: "As the agricultural export sector fell increasingly under the control of presidential cronies, the transparent use of presidential power to favor certain individuals and sectors of the economy destroyed the sense of community and common interest within the export sector."

Capital Flight
In Marcos's Philippines, as in so many other Third World countries, the beneficiaries of mercantile largesse seldom invested their easily amassed profits in local productive enterprises. More often they simply parked their capital in the United States or perhaps in Switzerland, countries perceived as providing better opportunities—as well as havens should their collective activities send their country into economic and social collapse. If Ferdinand and Imelda so assiduously exported their own ill-gotten gains, one could hardly expect their cronies to do otherwise. As

Bonner (1987:262) writes of the funds skimmed by top officials from the country's once lucrative coconut industry:

> Over the years the coconut planters, through the government im-
> posed levies, paid more than $1 billion into the bank. Where it went
> nobody knows—nobody, that is, except Marcos, Cojuangco, and
> Enrile. But there can be little doubt that it went into their osten-
> tatious life-styles. Cojuangco . . . owned . . . a $20 million stud ranch
> in Australia. Enrile . . . purchased two condominiums in a luxurious
> building on Broadway in San Francisco. . . . Enrile's wife, Christina,
> was the registered owner, and in 1982, when the Enriles purchased a
> $1.9 million apartment one block away, the two other apartments
> were transferred to a corporation called Renatsac, which was the
> backward spelling of Christina Enrile's maiden name.

The phenomenon of Third World elites transferring their money to the First World—capital flight as it is generally called—remains an under appreciated cause of economic stagnation. It represents a massive hemorrhage of capital, the lifeblood of genuine development. This flow of money from the poor to the rich countries probably outweighs the counterflow of aid and loans, and it may well form the largest net drain on Third World economic resources.

Capital flight represents, to a certain extent, the failure of Third World business classes to fulfill their historic mission to accumulate and invest locally. A Peruvian businessman depositing money in Miami may be acting in his own self-interest, but he is defying the needs of his society. Successful Third World countries have been able to retain capital during the crucial early stages of economic development. In South Korea, for example, capital flight has been virtually nonexistent (Fajnzylber 1990: 343). Fortunately, a few Third World leaders are beginning to realize the magnitude of the problem. Argentina's Carlos Menem, for example, must betray his own Perònista heritage in attempting to attract First World investment, but he argues that this is necessary because local capitalists have simply exported most of their funds (*Business Week*, September 24, 1990, pp. 60–61). But as the Philippine example shows, the champion capital exporters are often not business persons at all, but rather political elites who manipulate the market economy for their own personal advantage.

As Eric Jones (1988) shows, there was nothing magical about the capitalist transformation that brought wealth and power to the West. All societies, he argues, have a propensity for intensive economic growth. Throughout history, however, thriving economies have been strangled by elites mindful only of their own self-interests, anxious to seek rent rather than to invest in productive enterprises. In most of the Third World rent-seeking regimes are currently in power; some are so rapacious that they may accurately be labeled kleptocracies. It is thus hardly surprising that capitalism in this distorted form has failed to bring about global economic development.

The Newly Industrialized Countries (NICs) of Asia

A few Third World countries, however, have been able to avoid the traps described above and graduate to (near) First World status. Significantly, South Korea, Taiwan, Singapore, and Hong Kong have all found success in a capitalist rather than a socialist developmental path. At present, two other Asian countries seem poised to gain NIC status in the near future: Malaysia and Thailand (with Indonesia forming a less likely third). In examining these success stories it is important to note both the differences and the commonalities that have marked their economic policies. All have relied on the market, but most have found it necessary for the state to intervene heavily in crucial areas.

The only NIC that would please the heart of a radical free-marketer is Hong Kong, which probably embodies the doctrine of laissez-faire more fully than any other polity in the world (Bauer 1981). But Hong Kong is a poor exemplar, being a city rather than a country and a territory rather than a state. Moreover, Hong Kong has benefited greatly from its unique position as gateway to China.

Much more compelling is the fact that the largest NICs, Taiwan and South Korea, have consistently eschewed the laissez-faire doctrine. In both countries the state has worked closely with private enterprise, financial markets have been tightly regulated, and domestic industries have been protected from foreign competition during their critical periods of initial growth. In fact, in all of the newly industrialized *countries*, "state-led industrialization has become the norm" (Gereffi 1990:23). From its very beginnings, South Korean industrialization was premised on the notion of "guided capitalism" (Haggard 1990:68), with the state using its powers to sustain rather than to repress the market (Yun-Shik 1991:108).

Perhaps most importantly, Taiwan and South Korea for many years carefully guarded their supply of capital. South Korea's first economic boom, in fact, was coincident with a governmental clean-up campaign that "attacked conspicuous consumption by urban industrialists" (Haggard 1990:72). Until recently, South Koreans could travel overseas only for business or political purposes, since the state feared the hemorrhaging of capital that foreign travel would entail. Equally essential, the citizens of both countries took up the challenge themselves; in 1987 South Koreans saved 32 percent of their income. This was the second highest savings rate in the world, with only the Taiwanese coming out ahead (Porter 1990:467).

These policies served both Taiwan and South Korea well during their early years of phenomenal growth. Current trends, however, indicate that as they graduate to fully industrialized status, market-distorting restrictions are beginning to have deleterious effects. As a consequence, Schive (1990:289) strongly recommends economic liberalization, and there are some signs that this is beginning to occur in earnest. Both countries are also undergoing much needed political liberalization, although Taiwan has yet done little in the crucial area of democratization. The transition from the status of authoritarian, export-oriented industrializers to that of democratic, mature, industrial powers will not be easy, and both Taiwan and South Korea have seen their stellar rates of economic expansion falter over the past few years. Both countries have also experienced wage hikes that outpace growth, a trend worrisome to exporters, but hopeful for those who would welcome the emergence of broad-based economies that rely more on innovation than on cheap labor (see *Business Week*, December 3, 1990, p. 56; and *The Economist*, October 9, 1990, p. 33).

A consistent set of social characteristics also marks the successful industrializers. Most significantly, all enjoy relatively small income disparities. In 1980 the bottom 40 percent of Taiwanese citizens held a 22.7 percent income share, whereas in Mexico (in 1977) the bottom 40 percent held only an 8.2 percent income share (Haggard 1990:229). Moreover, through the 1950s, 1960s, and 1970s, Taiwan's wage structure grew increasingly egalitarian, while Mexico's grew increasingly inegalitarian (Haggard 1990:229). Indeed, in all measures of quality of life, the Asian NICs have shown remarkable progress, much in contrast to the stuggling Latin American industrializers.

Many observers have credited thoroughgoing land reform in South Korea and Taiwan, as well as in Japan, for laying the social and economic foundations for subsequent industrialization (see Haggard 1990; Gereffi and Wyman 1990). Postwar land redistribution created a strong stratum of mid-level peasants who were subsequently able to invest in agricultural production. As Haggard (1990:232) writes, "Taiwan's record, the best among Asian NICs, suggests the importance of improving the distribution of income *within* the agricultural sector." East Asian land reform was also vital for the creation of industrial policy because it demolished the notoriously conservative class of landed elites who would have otherwise opposed it (Haggard 1990:97).

All of the successful Asian industrializers have also invested heavily in human resources. In so doing they have created relatively healthy and remarkably well-educated populations. Michael Porter (1990:465) reports that the South Korean commitment to education is stronger than that of any other country he studied, a group including Germany, Sweden, and Japan. Significantly, the less successful Latin American NICS—Brazil and Mexico—have shown little interest in broad-based educational programs (Ellison and Gereffi 1990:386).

Finally, one must note that the Asian NICS are ethnically rather homogeneous. Such commonality has enhanced the emergence of a strong sense of national purpose—an essential precondition for successful development. The creation of a national bond is vital, Ronald Dore (1990: 361) reminds us, because it affects "the likelihood that the policy of the government will be accepted as legitimate—that it will be believed to be in the *national* interest, and not some conspiracy to benefit a sectional interest group." The absence of such unity may undermine development in other candidates for NIC-hood; tension between Malays and Chinese in Malaysia, for example, has already proven a strong fetter to economic growth. Polyethnic countries face the tremendous challenge of maintaining a sense of political and economic unity while simultaneously respecting cultural diversity and local autonomy. Despite fervent claims to the contrary, such a balancing act has seldom if ever been successfully achieved.

The differences among the various Asian NICS are, in the end, as instructive as their similarities. Only South Korea has fostered the growth of massive conglomerates, but its top four *chaebol* accounted for a remarkable 32 percent of its total exports in 1988 (Porter 1990:472). Taiwan's economy, in contrast, is dominated by a multitude of small, family-owned firms. Both systems offer certain advantages. Korean companies can more easily expand into expensive, high-tech frontiers, whereas Taiwanese firms prove highly flexible in the face of quickly changing demand patterns. Singapore presents another permutation, as it has prospered in large part by attracting subsidiaries of American and Japanese corporations. Indeed, Singapore has at times actually favored foreign over domestic capital (Haggard 1990). Thailand appears to be following Singapore's course, although in this instance few of the companies involved are American.[5]

Such disparities demonstrate that there is no single path to genuine development. Whether South Korea's or Taiwan's model proves more successful in the long run remains to be seen. It is clear, however, that in East Asia, just as in the United States, small companies have done less to control pollution, and are more difficult to regulate, than large companies. This is one of the reasons why Taiwan in particular suffers such horrendous environmental degradation (see *The Economist*, October 6, 1990, p. 20).

Despite such differences, it is important to remember that the Asian NICs share some significant general characteristics. All have embarked on a globally oriented program of export-led growth. At the same time, all have also exhibited a striking sense of common purpose, and all have discouraged the individual aggrandizement that comes at the expense of national development. Again, we can see that competitive cooperation (or, if one prefers, cooperative competition), proves to be a key to economic success.

Radical scholars sometimes dismiss the Asian NICs by claiming that the entire world cannot find prosperity by exporting cheap toys to the United States. Although this betrays a naive conception of the South Korean and Taiwanese economies, it does bring up a valid point. Development in the Asian NICs has depended crucially on massive exports to the industrialized world, especially to the United States. Both economic and political factors, however, limit the quantity of imported goods wealthy countries can absorb. Considering its debt burden and trade deficit, the United States cannot remain the mass market for all of the world's would-be export-oriented industrializers.

But the fall of the United States from global economic dominance may allow other countries partially to assume the role as export absorber. A rising society characteristically passes through several distinct stages, culminating—if it ascends to the position of global pivot—with a massive conversion to the doctrine of free trade. In fact, Japan has already reached the stage at which it must begin to open itself to imports; if it refuses, economic imbalances may well topple the entire world economy. Japan is indeed beginning to move in this direction. A significant proportion of the burgeoning output of Thailand, for example, flows to Japan, although it is notable that the great bulk of these exports are produced by Japanese corporations.

The economic evolution of Taiwan and South Korea lends credence to the notion that other countries might follow the Asian path of development. The now industrialized Asian nations began their ascents by exporting cheap consumer items, especially textiles. Subsequently, all have begun to graduate to more sophisticated products (Schive 1990). Such a move is indeed forced by the wage-level increases that accompany successful industrialization, but forward-looking governments recognize this as a healthy processes. The government of Singapore has, at times, actually encouraged wage hikes "in order to force firms to create more skilled positions and to adopt more capital- and technology-intensive processes" (Haggard 1990:146).

As successful industrializers abandon low-end products, poorer countries can step in to fill the void. Such a process helps explain the recently explosive growth of countries like Thailand and Indonesia. Development in these sub-NICs is also fueled by capital transfer, increasingly from the NICs themselves. In 1989, for example, Taiwan made direct investments in Southeast Asia worth more than $2 billion (*The Economist*, October 6, 1990, p. 33). But protectionist measures in the First World, most notably the Multifiber Agreement that limits textile imports, threaten the continuation of the process. As many nonradical environmentalists recognize, industrial protectionism presents one of the greatest obstacles to Third World development (Repetto 1990:16)—and to environmental protection as well.[6] As Porter and Brown write, "The liberalization of import restrictions on labor intensive manufactured goods . . . would encourage movement of capital out of resource depleting export crops . . . , thus easing pressure on natural resources" (1991:138).

In the end, successful industrialization requires the initiation of a virtuous spiral of capitalist development. Once it becomes clear that a national economy is quickly expanding, local elites will find it in their own best interest to invest at home. Foreign concerns too will smell profit, and capital will begin to flood to the region. Foreigners will, of course, repatriate much of their profits, but in a rapidly growing economy this will be of little consequence. Repatriation did not hurt the United States in the nineteenth century, and it is certainly not hurting Singapore today. Social conflicts will also diminish as joblessness eases and as health and education levels improve. All of these developments are visible today in Taiwan and South Korea. Moreover, as popular economic power grows, democratic forces cannot long be suppressed—again, witness South Korea. Of course, social tensions will never vanish (again, witness South Korea!); capitalist development, after all, never promises utopia.

As Fernando Fajnzylber (1990) brilliantly shows, Brazil and other stalled-out Latin American industrializers have pursued a policy of "showcase modernity" based on the emulation of U.S. consumption patterns among members of the elite class. Such elitist policies invariably fail. The genuinely developing nations of Asia have, in contrast, adhered to a Japanese model that stresses *relative* egalitarianism, constrained consumption, heightened investment, and broad-based education. In the end, such a recipe will prove essential not just for economic development, but for achieving environmental sustainability as well.

Environmental Conditions in Industrializing Countries
No radical environmentalist would find occasion to cheer the economic success of the NICS. Their growth has obviously come at the price of massive environmental degradation. Indeed, one would be hard pressed to find a city more polluted than Taipei outside of Eastern Europe. Burgeoning industries throughout the region spew out noxious wastes, and expansion-oriented governments have been unwilling to throttle their miraculous economies for the sake of environmental protection. Taiwanese companies, for example, generate some 19,655 tons of wastewater sludge every day, very little of which receives treatment (Chuang 1988:403). As Southeast Asia industrializes, many insidious forms of industrial pollution grow more threatening year by year. The Straits of Malacca, for example, are already highly contaminated with heavy metals (see Hungspreugs 1988; Jaynal 1985). In the East Asian NICS increasingly affluent citizens are consuming ever more resources, just as they are responsible for a burgeoning output of greenhouse gases. Intensive agriculture in Taiwan and South Korea relies on unsustainably heavy applications of fertilizer and pesticides, resulting in severe water degradation. Nor should one overlook South Korea's infamously efficient fishing fleet, which is devastating marine resources the world over.

Yet the NICS have made several impressive environmental strides. Most important is their precipitous drop in fertility. The South Korean population is now expanding at a rate of only some 1.2 percent a year (compared to 2.25 percent between 1965 and 1970), while Singapore has reached virtual demographic stability. Even Thailand has seen its annual rate of population growth decrease from 3.1 percent between 1965 and 1970, to 1.5 percent at present. Nondeveloping North Korea, in contrast, is still expanding at nearly 2.5 percent a year, while stagnant Bangladesh must accommodate some 2.7 percent more inhabitants every year. The Asian NICS are also preserving their remaining forests more successfully than are most of the economic laggards. South Korean and Taiwanese peasants, who find numerous opportunities in urban areas, are not forced to clear forest lands in order to construct marginal farm sites. More significantly, as prosperity builds, environmental movements are emerging with impressive alacrity. Taiwanese citizen groups have recently prevented the completion of several particularly noxious plants, and they have even forced one existing chemical factory to close (*The Economist*, September 8, 1990, p. 25).

Even the Taiwanese government increasingly recognizes that environmental degradation threatens future economic growth. Fortunately, it can afford to do something about it. In January 1991 Taiwanese officials announced an unparalleled $303 billion investment program targeting health, education, technology, infrastructure, and environmental protection (*Business Week*, March 25, 1991, p. 46ff.). Although environmentalists will regret that only $10.3 billion will be awarded primarily to environmental programs, one must admit that this is hardly a trivial sum. Meanwhile, Singapore has set its sights on providing environmental services to other Asian NICs. Singapore's Ministry of the Environment has even "set up its own company to sell expertise to neighbours" (*The Economist*, February 1, 1992, p. 80). Considering the problems at hand, and the economic resources available, environmental cleansing in East and Southeast Asia should present a substantial business opportunity.

The fear that multinational corporations will undermine Third World environments by using newly industrializing countries as pollution havens has also been partially allayed. First, only a few industries have experienced international redeployment due to environmental concerns. More importantly, successfully developing countries are becoming more sophisticated in bargaining with foreign firms, and they are increasingly unwilling to tolerate excessive pollution (Leonard 1988). Ironically, multinational corporations often operate under much stricter pollution control standards than do local firms (Pearson 1985:36–42; Pimenta 1987). Moreover, as Leonard (1988:212–13) discovered in his study of pollution in four industrializing countries, state-owned firms (especially in old industries) are usually the worst environmental offenders, while municipal governments are often not far behind.

Economic development ultimately proves to be a dual-edged sword. Although it brings about tremendous degradation, it also generates the economic resources necessary to begin solving environmental problems. In Asia environmental amelioration is already beginning to occur in the successfully industrializing countries. The air in Beijing and Calcutta, for example, is more seriously contaminated by particulates and sulfur dioxide than is the air of either Bangkok or Kuala Lumpur (*The Economist*, October 6, 1990, p. 19ff.). Ultimately, *lack of development* in the Third World is probably the gravest environmental threat the planet faces.

Southeast Asian Deforestation
Industrial pollution may be intensifying as Malaysia, Thailand, and Indonesia begin to develop, but forest destruction remains the region's most severe environmental problem. Southeast Asia's tropical rainforests (as well as its seasonally dry monsoon forests) are in dire threat of extirpation. Commercial logging, largely for export, has been a highly profitable endeavor for select concession holders, but any benefits that may have been conferred to the national economies are clearly unsustainable. Indeed, even the short-term profitability of Southeast Asian forestry has been vastly overrated (Repetto and Gillis 1988). Most cut-over lands are so degraded that forest regeneration would require hundreds of years. In fact, researchers at the World Resources Institute show that if such processes of natural capital consumption are factored in, Indonesia's recent annual GNP growth rate must be reduced from some 7 percent a year to some 4 percent a year (a figure that still remains substantially higher than its annual population growth rate of 1.6 percent) (*The Economist*, August 26, 1989, p. 53).

In stagnant, isolationist, and Buddhist-socialist Burma (Myanmar), by contrast, extensive forests have remained largely untouched. Anti-humanist greens, therefore might conclude that development after all is not worth the consequences. But it now appears likely that Burma too will sacrifice its trees. In 1990 the Burmese government decided to award most of its teak stands to Thai loggers, a move made in part so that Thailand would relinquish its support of the Karen and other autonomy-seeking nationalities within Burma's borders (*Far Eastern Economic Review*, February 22, 1990, p. 16ff.). A country that retains its forests not by virtue but rather by default—and which is happy to sacrifice them so that it might more effectively repress its own people—can hardly be considered an environmental model.

The deforestation of Southeast Asia is a global tragedy. One can only hope that successful industrialization, now only in its infancy, will soon begin to undermine the economic and political forces that lead to rampant logging. But more direct action is necessary if adequate natural habitat is to remain. First World countries should prohibit the importation of any tropical forest products harvested on a nonsustainable basis. Similarly, global environmental agencies should work with Southeast Asian countries to establish nature reserves large enough to ensure that healthy populations of elephants, orangutans, and rhinoceroses will be able to survive. By emphasizing tourism and allowing the sustainable harvest of select forest products, reserve planners might be able to support local economic development as well. Equally important, economists must strive to convince Third World governments that they have consistently undervalued "the continuing flow of benefits from intact natural forests" (Gillis and Repetto 1988:389).

But if nature reserves of adequate size to ensure minimum habitat protection are to be established and policed, Third World countries will have to forego certain economic benefits. Since the entire planet benefits from the preservation of natural diversity, reserve financing should ultimately be placed on a global rather than a national basis.

The Futility of Internal Colonization Schemes
Southeast Asia also forms a prime example of the environmentally disastrous and economically futile policy of attempting to alleviate rural poverty through population redistribution. On this score, at least, Arcadian and Promethean environmentalists agree fully. Several countries in the region have long encouraged peasants to move from high-density areas to frontier zones of sparse settlement. Typically, migration target areas are initially covered with thick forest and lightly inhabited by indigenous peoples—both of which retreat rapidly as the flood of immigrants arrives. The stated rationale behind most resettlement programs is both to relieve crowding in densely inhabited core areas and to spread the development process into previously underutilized hinterlands. An equally important hidden purpose, however, is to enhance state control over remote areas.

Such movements, whether state-funded or spontaneous, seldom if ever reduce poverty. Fertility rates generally remain high in the source areas, which in a few generations are more densely populated than before. Immigrants to the frontier zone, for their part, are usually confronted with poor soils and high disease rates, and they seldom have adequate capital to build viable agricultural enterprises (Fegan 1982). Moreover, migrants, desperate for the labor needed to clear land for production, often have very large families. Before long, even the former frontier zone may suffer from crowding and extreme poverty. In the end, genuine development must be characterized by intensive growth; the extensive growth characteristic of internal colonization is intrinsically destructive to the environment and does nothing to alleviate human suffering.

The Indonesian transmigration scheme, through which the central government has devoted millions of dollars to moving Javanese, Madurese, and Balinese peasants to the outer islands, has received scathing criticism from both environmental and human rights organizations (Reno n.d.). More often than not, transmigration has proven to be little more than a program of forest destruction, peasant immiseration, and tribal group dispossession. Such criticisms have provoked a major reassessment of the environmental consequences of relocation projects by the World Bank, a major transmigration supporter in the past. Whether the World Bank's newfound sense of environmental responsibility turns out to be genuine remains to be seen. Meanwhile, transmigration proceeds.

It is the Philippines, however, that illustrates the final results of internal colonization. At the time of the American conquest, several Philippine regions, most notably the Ilocano-speaking northwest coast of Luzon, suffered from rural overcrowding, but in general the archipelago was lightly peopled and huge expanses of land remained thickly forested. The American authorities, however, encouraged Ilocanos to move to Luzon's central plain, which was quickly denuded and converted to farmland. Soon the stream of migrants had to pushed into the vast Cagayan Valley to the northeast (McLennon 1980). Today, a few forested frontier zones persist in the Cagayan drainage, but they are in fast retreat. More significant is the fact that the Cagayan Valley has since become one of the Philippines' poorest and most lawless regions. The same process has been repeated elsewhere in the country, especially on the islands of Mindanao and Mindoro. Because of internal migration, small-scale societies have been destroyed and wildlife habitat has vanished. Today the only substantial remaining frontier is on the island of Palawan, but it too is undergoing rapid colonization.

With 65 million inhabitants expanding at an annual rate of 2.5 percent and suffering under an economic regime that shows no signs of sustained development, the Philippines exerts tremendous pressure for continued population dispersion. But soon the migration stream will come to an end, simply because the frontier will have vanished. One may well ask what the Philippines has to show for all of this. Its population continues to be impoverished—despite the bright prospects that it held only some thirty years ago. With its frontier safety valve operating for so many years, the Philippine government was able to avoid the fundamental socioeconomic restructuring that could have built the foundation for genuine, intensive development. Now, however, options are much more limited. The country is soon even going to have to begin importing lumber, as its own forests are virtually exhausted.

To avoid internal colonization, Third World countries can encourage *small-holder* agricultural intensification, much as did Japan, Taiwan, and South Korea in the postwar years. (Any program aimed at helping large-scale farmers will likely backfire; the mechanization of southern Brazilian farming may have resulted in an overall decline in Brazil's rural population [Bradley and Carter 1989:118], but it also led to a massive influx of dispossessed peasants into the rainforests of Rondonia.) Small-holder development, in turn, usually requires massive land reform. As it stands now, especially in Latin America, peasants are often forced to migrate simply because landed elites occupy, and underutilize, the best lands. In many cases, enactment of a more progressive land tax code alone might force ineffective elite landholders to divest themselves of their properties (Gillis and Repetto 1988:405). If the most fertile lands were distributed to peasants, output would increase, easing the pressure for migration.

In the short run, one must admit, agricultural intensification usually entails ecologically damaging chemical applications. Yet the overall environmental costs of chemical-intensive farming are lower than those of internal colonization. The social costs of modern intensification have also been overemphasized by radical scholars. Although the green revolution has been justly criticized on both social and environmental grounds, in many Asian villages small-scale cultivators have been able to use certain pieces of the green revolution package to their own benefit (Critchfield 1983).

Ultimately, however, a more environmentally benign form of high-yield agriculture accessible to a wide spectrum of peasant cultivators must be developed. This might rely on low-tech aspects of indigenous intensive farming (Richards 1983; Wilken 1987) combined with certain high-tech bio-engineering innovations. Indeed, Ghatak (1988) convincingly argues that a more environmentally benign second green revolution might occur in the tropics. This would be based on both ancient and modern biotechnologies, the latter including genetic engineering.

However important rural development is, it must still be joined to urban and industrial growth. A sufficient number of urban jobs must emerge to attract and support in decent fashion would-be land colonizers. Here again we can note the success of Taiwan and South Korea, and contrast it with the sad failure of the Philippines, and indeed most other Third World countries. Closely connected is the issue of population, for as industrialization proceeds, fertility rates generally decline. But as with other developmental issues, demography proves more complex than it might appear at first glance.

■ The Promethean Response to Population

Paul and Anne Ehrlich's (1990) recent work, *The Population Explosion*, is a necessary starting point for environmentalist demography. The Ehrlichs argue fervently against the population optimists, both those who welcome an ever increasing tide of humanity and those who assume that stabilization will automatically occur as poor countries pass through a second demographic transition. The Ehrlichs wisely regard both views as naive and dangerous. Increasing human numbers cannot help but translate into decreasing biotic diversity, while the global fertility decline over the past several decades is far too small to portend salvation.

The Ehrlichs attack the simple version of the demographic transition thesis for postulating a direct link between a given country's level of economic development and its rate of population growth. As they demonstrate, many areas of the world exhibit strikingly divergent patterns. We may elaborate this point by comparing levels of per capita GNP (admittedly, a poor measure of development, but the best one available) and rates of demographic expansion for select countries. In doing so we discover that Greece and Portugal have achieved virtual population stability, and may soon begin to decline, despite relatively modest levels of development (per capita GNP at 5,340 and 4,260 respectively), while Libya and Oman, with similar levels of per capita GNP (5,410 and 5,220 respectively), are expanding at a remarkable rate (4.2 percent and 4.7 percent respectively; all figures from the World Bank 1990). We might also note that several of the world's more impoverished areas, such as Sri Lanka and India's Kerala province, have seen substantial drops in their growth rates.

The forces behind such demographic patterns are far too complex to be encompassed by any single factor explanation, or even to be discussed adequately in a work of this nature. It is necessary to note, however, that the population explosion of the current era cannot be attributed merely to the decline in death rates that followed the diffusion of modern medical and sanitary practices. Far too many environmentalists accept this simplistic notion at face value. Accordingly, their demographic theories unrealistically ignore fertility rates.

Fertility patterns varied tremendously from society to society well before the transition to modern, industrial regimes. Historical demographers have, for example, uncovered a substantial increase in European birth rates in the immediate preindustrial period, a time marked by the spread of large-scale rural craft production. Such proto-industrial employment may have enabled peasant families to escape family size limitations previously imposed by the desire to avoid subdividing farm plots and by the need to delay marriage until adequate dowries had been accumulated (Kriedte 1983). Japan, however, shows a different pattern. In the second half of the Tokugawa period (Japan's long epoch of isolation), the country's population stabilized as families adopted strict control measures (including abortion and, at least in some cases, infanticide)—at precisely the same time as proto-industries developed (Hanley and Yamamura 1977). Only when Japan opened itself to the world economy in 1854 and began to industrialize did its fertility rate increase. In Southeast Asia the early modern period was similarly characterized by very low rates of population growth, a phenomenon definitely not attributable to high mortality rates. The "shift to rapid population growth in the nineteenth century," Anthony Reid (1987:43) concludes, was probably brought about by changing social organizational patterns coincident with the spread of Western power, and by the diffusion of lowland Southeast Asian cultural norms into upland areas.

A powerful school of materialist demography has recently argued that fertility varies according to the economic value that children confer to their parents (Caldwell 1978; Cain 1981). Simply stated, high birth rates may be expected where children labor diligently for their families and where they form their parents' sole source of social security. In many peasant societies a family's prosperity depends crucially on how many children (often, how many sons) toil on its behalf. In contrast, in the more developed societies where children require expensive education and generally fail to contribute to the family account, low birth rates may be expected.

This value of children hypothesis may help explain the rise in fertility that often accompanies the initial movement away from traditional social arrangements. In the early stages of development young people often find new opportunities to earn money (albeit in meager amounts) in the burgeoning cities. By finding employment in distant areas, children no longer form as much of a pressure on local resources, thus diminishing a previously existing fertility constraint. More importantly, by participating more fully in the commercial economy, children may be able to remit small amounts of cash to their parents, thereby bolstering the latters' economic standings.

In many respects, this preliminary stage of economic development is the most environmentally threatening. As agriculture is commercialized and chemicals introduced, degradation usually accelerates. Cities often grow more quickly than their local environments can tolerate, pollution thickens, and booming fertility results in rapidly mounting population pressure at the national level. If economic growth then stalls out, forestalling the emerging modern sector's ability to absorb the increasing human burden, an expanding class of landless rural inhabitants may be forced to clear new lands in marginal environments and in formerly rich wildlife habitats.[7]

The value of children hypothesis also helps explain the fertility decline experienced by successfully industrializing societies. As prosperity builds, other economic alternatives soon outweigh family remittances. More importantly, in an industrial society people begin to become individualized, losing intimate contact with their extended families and often even severing the economic bonds with their own children. As parental authority collapses, children may no longer be compelled to contribute to the family budget, while increasingly affluent parents will be more inclined to assist their children than to derive succor from them.

The fertility decline that usually accompanies urbanization and industrialization ought to present an unsettling paradox for eco-radicals. Their own envisioned future, one of nonindustrial population stabilization, could only be realized if all vestiges of development were removed, returning the entire world to a truly premodern mode of existence. Not only would modern transport systcms have to be dismantled, but all modern medical technologies also would have to be abandoned; a high death rate was, after all, a major contributor to the *relative* stability of most premodern population regimes. But if we admit that such a scenario is both fantastic and bigoted, we can only conclude that the eco-radical agenda would work strongly against population stabilization. By opposing urbanization and romanticizing the intimate surroundings of extended families and small communities, radical environmentalists struggle against the very forces that reduce fertility. Ultimately, by fighting against industrialization they would only help trap the Third World in the hyper-destructive state of initial modernization.

Beyond the Value of Children
Despite its manifest strengths, the value of children hypothesis cannot be taken as the last word in demographic theory. Many societies simply exhibit contrary patterns. In the Buguias region of the Cordillera of Northern Luzon, for example, the birth rate remains high, despite the fact that children form a net drain on parental resources. Young men even in their twenties commonly return home to ask for supplements when they find themselves in financial difficulties. Indeed, I was continually asked whether it was true that in the United States children are forced to support themselves upon reaching eighteen years of age. As one elderly man stated, "We wish we could do the same, but it is against our

ugali [culture]; in Buguias we must take care of our children no matter how old they are."

In the case of Buguias I was tempted to explain the persistently high birth rate on cultural rather than material factors. In the Buguias ideology a person's afterlife position depends on the animal sacrifices made by his or her descendants. The more progeny one has, the higher one's heavenly station is likely to be. As one elder phrased it, "if you have no children you are erased from the map of Buguias." The difficulty with this line of reasoning, however, is that members of the Christian minority, persons who deny such beliefs, have roughly the same fertility rate as the members of the Pagan majority.

Globally, it is notoriously difficult to ascribe fertility rates to religious beliefs. Catholic Italy, for example, has one of the world's lowest birth rates. Only in the case of religious groups that segregate themselves from larger national societies can one occasionally identify distinct patterns. American and Canadian Hutterites and other Anabaptists, for example, are famous for their remarkable fecundity. The clannish Mormons of Utah and southeastern Idaho also stand out strongly on maps indicating fertility rates.

In the end, we must conclude that a multitude of forces, some overt, other subtle, determine the fertility pattern of any given population. Other factors that cannot be analyzed here for lack of space include inheritance norms (especially whether land is divided or goes to a single heir), the presence of social institutions that promote celibacy, health and nutrition levels (healthy, well-fed populations can support high fertility rates; those with endemic venereal disease often have trouble reproducing at all), the work patterns of women (the meager body-fat reserves of highly mobile hunter-gatherers inhibit fertility), nursing patterns (prolonged lactation reduces fertility), and even the horse-riding, bathing, and undergarment-wearing habits of men (long hours in the saddle, hot baths, and tight underwear all reduce sperm motility). And last, but by no means least, we must take into account governmental population policy. Despite the claims of many radical theorists, population control programs often do succeed, at least to a limited extent. As Repetto (1985:135) argues, "Most countries that have experienced rapid fertility declines have made vigorous efforts to bring modern means of birth control within reach of the entire population."

One specific factor, however, does seem to be of overriding significance: the social position of women. Several studies have shown a strong correlation between the average levels of female education and overall fertility rates (Tienda 1984:163). In strongly patriarchal societies, men often want as many children (generally, sons) as possible, and in such a social environment they are usually able to overrule their wives, who may well hold different opinions. Such patriarchs typically exercise tremendous power over all of their dependents, and thus by having numerous offspring they can augment their own economic positions and enhance their own social prestige. Moreover, a man, unlike a woman, can generate dozens of children at no cost to his own body. Children may confer many benefits to their mother, but pregnancy, childbirth, and nursing are all enormously taxing. All other things being equal, women generally prefer smaller families in which each child can be afforded plentiful attention.[8]

The position of women thesis helps explain several anomalies in global demography. The patriarchal countries of the Middle East all have high fertility rates, but the richest among them, such as the United Arab Emirates (with a per capita GNP of $18,430), are actually expanding much more quickly than several of the much poorer ones, such as Egypt. In South Asia birth rates are moderately low in areas where women are relatively empowered, such as the Indian state of Kerala, whereas in zones of strong patriarchy and much higher levels of per capita wealth, such as the Punjab, birth rates are elevated.

Accordingly, the global dismantling of patriarchy must be seen as an essential precondition for reaching population stability. Unfortunately, this goal seems distant, in part because residents of the First World are so little concerned. We hear few reports of the shocking conditions in the patriarchal tyranny of Pakistan, a country in which a woman's testimony in court is regarded as worth exactly half of that of a man. Overall, the position of women in Pakistan is much like that of blacks in South Africa. But whereas racial apartheid has rightfully provoked global outrage, gender apartheid is all but ignored. Where are the campus protests against America's traditionally cozy military relationship with Pakistan? Where are the calls for sanctions?

This silence indicates that the deeply seated structures of patriarchy have not been excised from the American consciousness. But it also derives in part from one of the favored myths of the American left. According to contemporary leftist ideology, all of the Third World's problems are the result of Western imperialism and neo-imperialism. Such nonsense blinds radicals to the undeniable fact that patriarchy is not only indigenous to virtually the entire world, but that it has reached its apogee in certain non-Western cultures. Historically, European women may well have been second only to Southeast Asian women (on the latter score, see Reid 1988) in their overall degree of social power;

the societies of the Middle East and North Africa, South Asia, and East Asia have all been far more repressive. In fact, through most of the world, Westernization is equivalent to *partial* feminization—which is one reason why it is so detested by certain fundamentalist Islamic clerics.

Population Programs
In order to defuse the still ticking population bomb, the prime requirement is global development, both social and economic. Third World countries must invest in education, institute social security programs, and, most importantly, achieve equality of the sexes—and First World countries must donate substantial funds to such efforts. But it is questionable whether such fundamental social transformations could be accomplished quickly enough. State-sponsored population programs, based on subsidized birth control and on (voluntary) sterilization, are clearly in order. In the most densely inhabited Third World countries, however, even more forceful population control measures might be necessary. Perhaps China should be commended on this score, as reprehensible as the Chinese regime is in most other respects. Although the one child policy has been accompanied by some staggering human rights abuses, considering the magnitude of its population problem, China has few options. Unfortunately, environmental requirements occasionally conflict with human rights. In such cases, the survival of the planet must be granted priority.

Land and Population
Although virtually all Third World countries are growing at an alarming rate, it is important to recognize that many poor lands are still sparsely populated. Environmentalists, unfortunately, have seldom differentiated the demographic prospects of a country like Bangladesh (with 8,400 persons per 1,000 hectares) from those of a country like Gabon (with 44 persons per 1,000 hectares). Because of this oversight, anti-environmentalists can argue with some force that concerns about overpopulation are outrageously overblown.

There is a modicum of truth to this allegation. While most scholars would have a difficult time denying that Bangladesh, Java, Burundi, and the lower Nile Valley are overpopulated, one could hardly say the same for Bolivia, Angola, or Mongolia. By Asian standards, virtually all of sub-Saharan Africa (excluding Nigeria and the rift zone) is *sparsely* inhabited. Moreover, it is essential to recognize that food production (except in Africa!) has generally been keeping pace with population growth. In fact, famines are much less common than they were 100 years ago (Tarrant 1990:467).

Certain African leaders have, in fact, argued that their continent is underpopulated and hence unable to reap significant economies of scale. They trace this paucity of human numbers to the depredations of the slave trade, and they are indeed correct in noting that slavery resulted in a long period of African demographic stagnation. Similar patterns of historical depopulation at the hands of European imperialism may also be detected in much of Latin America, where several regions have still not recovered from the massive die-off that accompanied the introduction of European diseases in the sixteenth century.

Most environmentalists, not surprisingly, reject such reasoning out of hand. The Ehrlichs (1990) argue, for example, that Africa must be grossly overpopulated, despite its moderate population density, simply by virtue of the fact that it is increasingly unable to feed itself. They stress that much of the continent is cursed with poor soils and thus could never support the crowded settlements found in fertile Asian deltas. There is some truth here as well. Angola is no barren wasteland, but then again it is hardly an agricultural paradise. Only a few scattered African regions, most of them in the Rift Valley, are blessed with rich soils.

But this line of reasoning is also limited. While most African soils are poor, they can produce abundant crops if appropriate techniques are employed. In several densely populated areas of Nigeria, for example, local farmers have devised ingenious methods of coaxing good crops from meager soils (Netting 1968). Moreover, as the Ehrlichs recognize, the food crisis in contemporary Africa derives in large part not from intrinsic carrying capacity limitations, but rather from state policies that favor urban dwellers over peasant agriculturalists. Most African states have long mandated low food prices in order to allow low urban wages and to quell restive urban mobs. Faced with dismal grain prices, peasants are often unwilling to grow for the market, resulting in chronic production shortfalls. Although urbanization remains environmentally desirable, state policies that unfairly favor cities by undercutting agriculture will quickly prove destructive in societies struggling to produce adequate foodstuffs.

More compelling than fears about existing overpopulation in Africa, however, is the realization that, given current trends, severe demographic stress will not be long in coming. Thirty years ago Kenya was lightly populated, holding only some 6.3 million persons. Today its population stands at 25 million, and by 2025 it is expected to have reached 77.6 million. In other words, Kenya will have been transformed from a sparsely to a densely populated land in only some sixty-five years. As this process continues, most Kenyans will find their lives growing ever more precarious.

Another equally compelling reason for population stabilization even in lightly populated Third World countries stems from the requirements of the economic development process itself. Development can only occur if per capita economic activity increases, which means that economic growth must continually outpace population growth *regardless of population pressure*. It is quite a challenge, however, for any economy to grow more rapidly than 3 percent a year, the rate at which most African populations are currently expanding. Although the Asian NICs have easily maintained much higher rates of economic growth, most other countries, including the United States, have not been able consistently to accomplish as much for many years. Even many prosperous oil exporters are now de-developing insofar as their populations are expanding more rapidly than their economies. In sub-Saharan Africa, a few countries are tumbling downward in a de-development spiral. Zaire, for example, experienced a real per capita annual GNP growth rate of −1.6 percent between 1980 and 1989 (World Bank 1990:9).

A rapidly growing population places innumerable strains on all but the most vibrant economies. The infrastructural investments needed merely to accommodate the burgeoning numbers of young persons are staggering. Many poor countries are thus beginning to see declining levels of education, the Philippines being a prime example. Moreover, even successful educational programs can become counterproductive under such circumstances. The huge generational cohorts of youngsters found in fast-growing countries seldom find adequate employment when they graduate. The resulting contingent of educated but unemployed youngsters makes for social dynamite. In much of the Middle East, such dissatisfied young men are increasingly embracing an uncompromising, hyper-patriarchal form of Islamic fundamentalism. Such a movement bodes ill for both the global environment and for the alleviation of oppression within their own countries.

Finally, population stabilization throughout the Third World is essential for the maintenance of biotic diversity. Given present demographic trends, Africa's remaining havens for large mammals will give way to human pressures within several generations. At present, the best hope is that a few African countries will find it in their best interest to preserve sizable areas, if only to attract tourist dollars. Without economic incentives, habitat preservation will prove an expendable luxury in the increasingly desperate nondeveloping world. In the short run, engaging in eco-tourism is, in most cases, one of the best investments an affluent First World environmentalist can make.

Such are the demographic problems and prospects of the Third World. The task remains, however, to examine population patterns in the industrialized countries.

Population in the Developed World
Leftist critics have often dismissed concern about Third World population growth as little more than an imperialist ploy designed to allow the wealthy nations continued access to the bulk of the planet's resources. As the preceding discussion shows, denying the problem of demographic expansion is both naive and potentially destructive, but the point remains valid that pollution and resource depletion are in large part attributable to the world's wealthy societies. Any political program that would attempt to limit population growth in the Third World merely to maintain resource flows to the First World is indeed morally untenable.

But the discrepancy between the wealthy and the poor countries continues to present a special challenge for those who regard sustained economic growth as impossible. If one holding such a view wishes to avoid the taint of conservatism, he or she must advocate either a decline in the First World's living standards or a massive drop in its population. While most greens favor the former approach, Paul and Ann Ehrlich (1990) argue forcefully for the latter.

According to the Promethean environmental perspective, neither of these two alternatives is desirable. Although the Ehrlichs' recent work on the population explosion presents a valuable overview of Third World population problems and prospects, I must conclude that its analysis of demographic patterns in the First World leaves a great deal to be desired.

According to the Ehrlichs, a wealthy country like the United States, regardless of its population density or demographic trends, suffers from just as severe a population problem as an impoverished, closely settled, and quickly growing country like Bangladesh. America, they claim, is grossly overpopulated. Yet this assertion runs counter to common sense; ours is a country of moderate human density blessed with fertile soils and burdened by mounting agricultural surpluses. To take the Ehrlichs' next step and argue that the United States is undergoing a massive *population explosion*, brought on by an unsustainably high fertility rate, requires a concerted suspension of disbelief. The average American woman, after all, gives birth to only 1.9 children, a figure well below the replacement rate of 2.2. The Ehrlichs are able to accomplish such intellectual gymnastics, however, by mathematically equating affluence and technology with population. Although America's affluence has been static for some time, its technology is advancing, thereby forming, according to the model, the equivalent of a population explosion.

The essential question is whether such an equation is either scientifically meaningful or politically useful. Unfortunately for the Ehrlichs, the more carefully one scrutinizes the I=PAT model the more specious it is revealed to be. First, the equation is so vague as to be virtually nonsensical. A given area's population is a discrete number, and thus can easily be factored into any number of equations. Affluence, on the other hand, is a more slippery notion, being imperfectly quantifiable through such crude measures as per capita GNP. But even if it could be unproblematically measured, affluence would never provide an accurate indication of impact. Two equally affluent countries can be responsible for vastly different degrees of environmental degradation, depending on how their wealth is obtained and how their societies are organized. This is readily evident, for example, if one contrasts Sweden, wealthy yet relatively benign, with the United States.

Technology, the "T" factor of the equation, is far more problematic. To begin with, technology simply cannot be quantified, and is thus useless in an equation. If the technology of the United States were to count as 100, what then would be the value of Sweden's or Japan's—much less Russia's? Furthermore, although many technological advances are indeed ecologically disastrous, others—of equal sophistication—are environmental godsends. According to a strict reading of the Ehrlichs' model, every advance in pollution control technology (an increase in "T"), would contribute to the population explosion. Most advanced technologies, however, are environmentally ambivalent; the manufacture of computer and telecommunication equipment, for example, generates toxic waste, but, if put to good use, the resulting products can be environmentally beneficial. How then do such technologies fit into the equation? Ultimately, I=PAT is an example of pseudo-science, impressive only to those mystified by equations and other scholarly trappings.

The Ehrlichs' demographic thesis is also undermined by their implicit assumption that environmental degradation is a unitary phenomenon. According to their arithmetical reasoning, each society exerts a singular force on the environment, a quantity calculable as the figure "I." But in reality, environmental impact derives from such a wide array of human activities that quantification is impossible. Different human groups exert incommensurable kinds of environmental pressures. No single metric of impact can ever be derived.

In particular, wealthy and poor societies are characterized by markedly different patterns of environmental degradation. The average inhabitant of the First World is extremely taxing on the earth in many crucial areas, but in others may be completely innocent. Which country—for example, the United States or Yemen—exerts a greater destructive force on the earth's few remaining rhinoceros herds? It is essential for environmentalists to acknowledge the fact that habitat destruction is occurring more rapidly in the poor than in the wealthy countries. Of course, Third World peasants are not generally culpable, for they are usually forced to degrade their lands by circumstances beyond their control. But questions of blame notwithstanding, it remains true that relentlessly expanding Third World populations can hardly help but displace wildlife, while stabilized or shrinking First World populations can increasingly afford to return land to wildlife habitat, if only they would choose to do so.

The absurdity of regarding impact as a unitary phenomenon is evident in the Ehrlichs' contention that a single American generates the same degree of environmental destruction as do 140 citizens of Bangladesh. In accordance with such reasoning, Bangladesh's human impact would only be as large as that of the United States if Bangladesh supported a population density 140 times greater than that of the United States, or, in other words, if Bangladesh held some 9,240 persons per square mile for a total population of 513,725,520 (instead of the 115,000,000 it contains today). It is difficult to imagine provisioning a human community in such a densely populated environment, let alone preserving any biotic diversity. The Ehrlichs' impact equation may be heuristically employed for select global issues, such as carbon dioxide production, but as an all-encompassing measure of environmental degradation it has no meaning whatsoever.

Considering the fact that the United States is not heavily populated and has reached a native state of population stability if not decline (its growth being attributable solely to immigration), one must conclude that it is not experiencing a population explosion. It does, of course, suffer from many staggering environmental problems, but these stem from how its population lives and the technologies it employs—and elects not to employ.

The Ehrlichs, however, dismiss the contention that the American environmental crisis derives from behaviors and policies rather than mere numbers. To their way of thinking, we must accept the realities of the American lifestyle: "To say that [the United States is not overpopulated] because, if people changed their ways, overpopulation might be eliminated, is simply wrong—overpopulation is defined by animals that occupy the turf, behaving as they naturally behave, *not by a hypothetical group that might be substituted for them*" (Ehrlich and Ehrlich 1990:40).

This is an unusual argument indeed for two environmentalists to make; most of their colleagues, after all, argue that Americans behave in a distinctly *unnatural* fashion. Actually, human behavior can never be adequately conceptualized as either natural or unnatural; unlike most other animals, our actions are shaped by cultural norms, malleable institutional guidelines and barriers, and personal initiatives. The Ehrlichs' argument is also self-canceling, for the same logic would compel us to conclude that reducing our population would be equally impossible, since it would require "substituting a hypothetical population" with a different fertility pattern "for the one presently occupying the turf."

One might reasonably argue that the United States does has a population problem, however, if one considers migration. Owing to immigration, the American population is expanding at roughly 1 percent a year, a rapid pace by global historical standards. Certainly this level of growth can be accommodated for some time, but if it persists the environmental consequences will eventually become severe. Yet in many ways immigrants help create a more vibrant and productive economy, a contribution that has notable environmental benefits. American migration policy is obviously a matter for careful and reasoned debate within the environmental community.

American Population Reconsidered
The Ehrlichs' horror at the thought of an American woman giving birth to more than two babies probably stems either from a profound distaste for the human species or from an attempt to mollify leftist opinion (which automatically suspects any concern with Third World population as a mask for racism). In either case, the arguments they present are counterproductive for the environmental movement. The occasional misanthropy of green activists has long been the movements' greatest stumbling block in gaining widespread acceptance; the louder one denounces humanity, the less public support one may expect. If, on the other hand, the Ehrlichs are merely trying to curry favor on the extreme left, I believe they are committing a grave tactical error. While leftists may hold great power within universities, in the larger scheme of American politics their influence is nil.

By adhering to a radical stance of baby-bashing, the Ehrlichs risk reducing the very real problem of global population pressure into an easily derided crank theory. In regard to population more than anything else, environmentalism must be based on a solid theoretical and empirical foundation.

Conclusion

To sum up, the notions of global development and population growth prevalent in the eco-radical literature are no more helpful for constructing an efficacious environmental movement than are the eco-radical theories regarding technology and economic scale. Most radical greens walk on shaky ground indeed when they enter the Third World, although little do they realize it. They want desperately to alleviate human misery, but they can only advocate programs that would preclude genuine development. They hope for nothing more than the stabilization of population, but they can only propose policies that would lead to continued growth. They are horrified by wildlife destruction, political repression, and the exploitation of women and minority groups, but they hurry to absolve Third World societies of *all* responsibility, and in so doing obscure many of the reasons why such problems exist in the first place.

To understand why eco-radicals so misunderstand the global predicament it is necessary to examine in greater detail their romantic sensibilities. The conclusion of this work will therefore open with a brief return to the thesis of primal purity.

Conclusion

*The rocky crests, the juices in
the meadows, the body heat of
the pony, and man—all belong to
the same family. Chief Seattle
of the Suquamish tribe.*

Orientalism and Radical Environmentalism

Chief Seattle's paean to nature occupies a central place in contemporary environmental literature. His speech strikingly reveals the unflattering contrast between Euro-American attitudes toward nature, based on thoughtless exploitation, and Native American attitudes, founded on spiritual union. Eco-radicals, in particular, find comfort in his words, seeing in them an eloquent demonstration of the superiority of the primal way and an unimpeachable accusation of Euro-American cupidity and destructiveness.

Unfortunately, the speech is fraudulent. These words were written not by a despondent indigenous philosopher in the days of his peoples' dispossession, but rather by a white American screenwriter, working for the Southern Baptist Convention (*Environmental Ethics* 1989:195–96). The attitude that they so eloquently express is not that of an indigenous people, living in harmony with nature until the arrival of rapacious Westerners. Rather, they exemplify the ideas that modern American society considers fitting for the less technologically oriented peoples who supposedly retain a primordial bond with the earth.

The Seattle hoax was uncovered by scholars sympathetic to eco-romanticism, and their exposé was printed in a journal of decidedly eco-radical inclinations. No doubt they were distressed to uncover the text's true authorship, but they realized how damaging the revelation could be if delivered by their opponents, and they sought to defuse any potential backlash by exposing the story themselves. Moreover, they argued that one fraudulent text would do little damage to the environmental move-ment's faith in Native American wisdom; just because Chief Seattle never uttered these oft-quoted words does not invalidate the many other statements expressing the indigenous American philosophy of nature.

The editors of *Environmental Ethics* are right to argue that ecological wisdom can be found in Native American philosophy (or, as I would prefer, philosophies). But this does not mean that we should simply dismiss the Seattle episode as an unfortunate result of sloppy scholarship conducted by overly enthusiastic environmental advocates. The speech is, after all, probably the most widely cited evidence for the superiority of American Indian traditions over those of the materialistic and ex-ploitative West. Seattle's (supposed) words obviously struck a chord not just in the minds of eco-radicals, but in those of the American public at large.

The appeal of Chief Seattle's speech, I would propose, lies in the longing for a return that invariably accompanies technological advance. Human beings usually crave progress, but rarely without misgivings. Such doubts are translated into visions of Arcadian serenity. As John Sisk (1991:239) shows, the philosophical roots of eco-radicalism can easily be traced to ancient Greek thought: "One thing is clear, however: the dis-content and disgust with civilization that we find in the Cynics and some of the Stoics are thoroughly familiar to us. They believed that after a promising and perhaps Hyperborean beginning in golden times, things had gone wrong as humans departed perversely from the way of nature. In due time this realization produced an immense amount of literature, ranging from elegiac regret to that despair of human reason and inven-tiveness always lurking in the dark shadows of romanticism."

Marianna Torgovnick (1990) has admirably chronicled the twentieth century's idealization of the primal past in *Gone Primitive: Savage Intel-lects, Modern Lives*. Our literature, art, and psychological theories—whether devised by leftists, rightists, or centrists—create an image of an idealized primitive. Fictitious notions of savagery conjure whole literary genres, while art and myths created by peoples deemed close to nature are appropriated and imbued with meanings of our own making. Such meanings, more often than not, prove foreign to the cultures whose forms and artifacts we preempt. Where modernist artists saw a frenzied sexuality in African masks, the artists who carved them saw rather a pleasing composure that brought to mind ritual activities of a decidedly nonerotic nature. Yet we continue to congratulate ourselves in making such misrepresentations, believing that we accord respect to primal peoples by elevating them to the status of nonalienated humanity in its essence, at one with earth and nature. In reality, this procedure denies them their own existence, making them instead a dumping ground for our own fears and longings, a "distorted mirror of the western self" (Torgovnick 1990:153).

Eco-radicalism carries this intellectual error beyond merely distorting primal peoples for present purposes. At the very foundation of eco-radical thought lies a gross distortion of human history, a singling out of the West as the sole source of environmental degradation, and indeed, in the most extreme examples, as the single repository of human evil. In doing so, it ironically perpetuates the discredited intellectual tradition of orientalism. Orientalists of past generations believed that the timeless East, a realm extending from Istanbul to Tokyo, was fundamentally homogeneous, at least when compared to the restless and dynamic West (the non-Eurasian world, in this scheme, was hardly considered worthy of discussion at all). Contemporary eco-radicals largely agree with the former contention, except that they reverse the moral signs: the soul of wisdom is now to be found in non-Western stasis.[1] For Roderick Nash (1989:113), foremost historian of the environmental movement: "The oriental mind tended to regard nature as imbued with divinity. . . . All beings and things, animate or inanimate, were thought to be permeated with divine power or spirit, such as Tao or, in Shinto, Kami." These sentiments fit remarkably well within the tradition of orientalism. Romantic orientalists have always hoped that the spiritualism of the East could "defeat the materialism and mechanism (and republicanism) of Occidental culture" (Said 1978:115). Even the more moderate environmental works often fall into the same mold. Philip Hurst, in *Rainforest Politics* (1990), for example, informs us that the traditional Eastern land management concept was based on communal control informed by holistic philosophies (1990:246–47), and he further implies that economic growth is strictly a "Western ideal" (p. 254).

As I have endeavored to show, the West holds no monopoly on environmental destructiveness. In fact, in many respects the East is guilty of the modern world's most extreme violations. Hong Kong, after all, is the center of the global trade in endangered species. The Chinese taste for bear paws and gall bladders, monkey brains and snake skins, elephant tusks and cat pelts is entirely indigenous—a fact conveniently ignored by the scores of environmental articles extolling the ecological virtues of Taoism and Buddhism, and informing us that the materialism (or, in some versions, the biblical tradition [White 1967]) of the West lies at the heart of the human assault on the global ecosystem.

More fundamentally, however, the very notion of the West is gradually

losing its meaning. The concept is flagrantly Eurasia-centric in conception, and has long been without a clear geographical referent in any case.
The rise of Eastern Eurasia is now demolishing even the socioeconomic
vestiges of meaning in the category of the West as well. Can we really
speak of a "West" that increasingly looks to Japan for technological and
financial leadership? It would appear rather that we are witnessing the
emergence of a global cosmopolitan culture, only some of whose roots lie
in Western Europe.

The academic left is now engaged in an all-out assault on the Eurocentric traditions of the university. The initial premise of this attack
is entirely legitimate; the traditional college curriculum vastly over-
rates Europe's importance. In fact, it is our staggering ignorance of non-
Western cultures that allows us to be so easily beguiled by eco-romantic
fantasies. But the left's usual methods of rectifying Eurocentrism paradoxically aggravate the central problem. Much radical scholarship actually retains a strongly Eurocentric bias, albeit one that disparages rather
than celebrates Western achievements. The non-European world increasingly receives attention, but only in its role as victim, or resister, of
Western exploitation. Little interest is shown for the historical development of Asian, African, and indigenous American cultural and intellectual traditions in their own right. The call is for the study of contemporary political works written by the downtrodden, certainly not for
the Analects of Confucius or Meso-American cosmology and calendrics
(D'Souza 1991). Scholars in anthropology and geography are increasingly
retreating from overseas fieldwork, which some see as inescapably impe-
rialistic, in favor of producing ideological critiques of earlier scholarship,
or of further dissecting the dominant tradition of the United States. In
the process, any hopes of establishing even partial cross-cultural understanding are relinquished.

The counterhegemomy of the academic left also proves intellectually
destructive in ignoring, if not denying, the various evils, both social and
environmental, that have always infected non-Western cultures. Most
discussions of slavery, for example, focus exclusively on the Atlantic
trade, overlooking the equally brutal export of Africans to the Middle
East. Orientalists, in both their traditional and modern guises, have
typically downplayed or apologized for Eastern slavery; some have even
regarded it as positive in crucial respects. But anyone who examines the
historical records of the East African slave trade, replete with such
horrors as the eunuch factories of upper Egypt, would have a difficult
time justifying such an attitude. As Bernard Lewis (1990) concludes in
his book *Race and Slavery in the Middle East*:

The myth of Islamic racial innocence was a Western creation and served a Western purpose. Not for the first time, a mythologized and idealized Islam provided a stick with which to chastise Western failings [p. 101].

The white man's burden in Kipling's sense—the Westerner's responsibility for the peoples over whom he ruled—has long since been cast off and seized by others. But there are those who still insist on maintaining it—this time as a burden not of power but of guilt, an insistence on responsibility for the world and its ills that is as arrogant and as unjustified as the claims of our imperial predecessors [p. 102].

I am not proposing that we ignore the brutality of Western societies, nor am I advocating a retreat into non-Western antiquarianism. Rather, I am calling for a globalist approach, one that takes seriously the historical-geographical unfolding of the entire world. All cultures should be studied, both in their internal dynamics and in their interactions with other cultural systems. Contributions to human culture have come—and will come—from all reaches of the world; no region should ever be denied or overlooked—or deemed primary. But equally important, we must realize that destructiveness and exploitation have been features of *all* human societies. To reduce the evil in human history to the functioning of a specific socioeconomic form (capitalism), or to limit it to a single subcontinent (Europe), is a foolish gambit indeed.

Radicals are correct to argue that we must pay particular attention to Euro-American imperialism, which completely redrew the map of the earth to its own advantage. But let us not ignore other imperialisms. In the modern world it is hard to find a country more brutally imperial than China, as the Tibetans can surely testify. Historically as well, the global comparative study of empire is essential for understanding the failings of human social and economic development. The Mongol conquests were not only some of the most vicious the earth has ever known, but they may have actually contributed to the rise of Europe—the only Eurasian civilization largely spared their depredations (A. Lewis 1988). Central Asian conquerors were also guilty of extraordinary environmental degradation, and some areas they plundered have yet to recover. As Grousset (1970:428, 429) explains:

At Zaranj, capital of Seistan, Tamerlane 'put the inhabitants to death, men and women, young and old, from centenarians to infants in the cradle.' Above all, Tamerlane destroyed the irrigation system of the Seistan countryside, which reverted to desert. . . . The desolation that strikes the traveler in this region even today is the result of these acts of destruction and massacre. The Timurid chiefs were finishing what the Jenghiz-Khanite Mongols had begun. Both . . . made themselves the active agents in this 'Saharifying' process, to which the center of Asia, by its geographic evolution, is already too prone. By . . . turning the land into steppe, they were unconscious collaborators in the death of the soil.

The Threat of Radical Environmentalism

Eco-radicalism is admittedly a marginal social movement, its adherents forming an exiguous ideological minority. One might be tempted to conclude that it poses no threat to our economy, our society, and our environment. This may ultimately prove true, but it cannot be assumed. Radical environmentalism enjoys substantial, and growing, intellectual clout. If its concerns merge with those of the broader academic left, a trend visible in the rise of both eco-marxism and of a self-proclaimed subversive postmodernism, we may well see the intellectual hardening of uncompromisingly radical doctrines of social and ecological salvation.

Academic radicals hope to create a potentially revolutionary intelligentsia in the United States. On this score they seem to be succeeding. But this intelligentsia, I believe, will find itself restricted to academia. The professoriate may be increasingly radicalized, and it may be able to recruit enough graduate students to carry on the cause, but its influence on the vast bulk of undergraduates will remain minimal. Despite receiving an increasingly leftist education, young college graduates, like their less-educated generational peers, are notoriously conservative. Indeed, voters born after 1960 are "by far the most Republican-leaning youths in the sixty year history of age-based polling" (Strauss and Howe 1991:326). The radical message does not appear to be sticking; quite the contrary, it seems to be backfiring—and with some heat.

A majority of those born between 1960 and 1980 seem to tend toward cynicism, and we can thus hardly expect them to be converted en masse to radical doctrines of social and environmental salvation by a few committed thinkers. It is actually possible that a radical education may make them even more cynical than they already are. While their professors may find the extreme relativism of subversive postmodernism bracingly liberating, many of today's students may embrace only the new creed's rejection of the past. Stripped of leftist social concerns, radical postmodernism's contempt for established social and political philosophy—indeed, its contempt for liberalism—may well lead to right-wing totalitarianism. When cynical, right-leaning students are taught that democracy is a sham and that all meaning derives from power, they are being schooled in fascism, regardless of their instructors' intentions.

According to sociologist Jeffrey Goldfarb (1991), cynicism is the hall-mark—and main defect—of the current age. He persuasively argues that cynicism's roots lie in failed left- and right-wing ideologies—systems of thought that deductively connect "a simple rationalized absolute truth . . . to a totalized set of political actions and policies" (1991:82). Although most eco-radicals are anything but cynical when they imagine a "green future," they do take a cynical turn when contemplating the present political order. The dual cynical-ideological mode represents nothing less than the death of liberalism and of reform. Its dangers are eloquently spelled out by Goldfarb (1991:9): "When one thinks ideo-logically and acts ideologically, opponents become enemies to be van-quished, political compromise becomes a kind of immorality, and con-stitutional refinements become inconvenient niceties."

But unlike the youth of today, few eco-radicals are cynics at heart. Quite the contrary, they remain intense idealists. Most radical environ-mentalists are member of the baby boom generation, which Strauss and Howe (1991) convincingly portray as inherently idealistic. Members of this cohort incline toward stern doctrines that picture the world in stark terms of good and evil. Most eco-radicals unambiguously define as good the realms of nature and primal culture, and as evil the domain of modern industrial society. Marxists, whether environmentally oriented or not, similarly draw strength from moral absolutes, and some are honest enough to admit that they consider capitalism to be nothing less than evil (Walker 1989:160). Even the supposedly skeptical relativists of the subversive postmodernist camp often follow suit, finding within the dominant culture all attributes commonly associated with evil.

Throughout this work I have referred obliquely to the religious aspects of eco-radicalism; in concluding I can hardly emphasize them strongly enough. From its beginnings the contemporary environmental move-ment has been obsessed with religion (for example, Barbour 1973). In-deed, Scheffer (1991:7) describes environmentalism itself as a kind of religious reformation. While relatively few eco-radicals have formed full-fledged churches (see, however, Gelber and Cook 1990), all have a funda-mentally religious, indeed, millennarian, outlook on the world. To the true believer, the modern world is thoroughly derelict, and it will either perish for its sins or we will collectively find eco-salvation. Many green radicals are also strongly attracted to asceticism, again showing their religious zeal.

While I have nothing against religion in general or environmental religion in particular, I do fear the religious intensity that so often infects members of an idealistic generation. In the heat of ideological fervor, true believers have time and again proved themselves capable of committing dreadful acts in the name of a higher good. While seeking moral and social perfection, those committed to a purist vision consistently work against the development of the social consensus necessary to make reforms work. Whereas social progress demands broad inclusion, radicalism excludes all persons judged sinful—or, in the current jargon, politically incorrect. Where workable solutions to social and environmental problems require compromise, radicalism calls for implementing only one's own program while vanquishing those of one's rivals.

This quasi-religious character of the radical environmental movement draws on its great strength: a consistently utopian vision. Imagining a world in which human beings live in harmony with each other and with nature is a rewarding and comforting exercise, and the utopian imagination deserves credit for enriching global culture. By selecting specific social traits and environmental relations from a host of small-scale societies, and by placing the idealized melange in a timeless land unthreatened by rapacious neighbors, deep ecologists have indeed constructed a tempting scene, an ideal escape from the frenetic world we all inhabit.

But for all of its attractions, utopia remains, and will always remain, "no place." Although the vision is easy to conjure, the reality is elusive. In fact, those political regimes that have struggled hardest to realize utopian plans have created some of the world's most dystopian realities. Unfortunately, Americans as a people seem uniquely drawn to such fantasies, and a right-wing variant of utopianism has even guided our recent national administrations; as Robert Kuttner (1991:5, 157) shows, laissez-faire itself is an ideologically driven utopian scheme that has dire consequences for the earth's economy and ecology. As Michael Pollan (1991:188) eloquently demonstrates, eco-radicalism and right-wing economic theory are more closely allied than one might suspect: "Indeed, the wilderness ethic and laissez-faire economics, as antithetical as they might first appear, are really mirror images of one another, Each proposes a quasi-divine force—Nature, the Market—that, left to its own devices, somehow knows what's best for a place, Nature and the Market are both self-regulating, guided by an invisible hand. Worshippers of either share a deep, Puritan distrust of man, taking it on faith that human tinkering with the natural or economic order can only pervert it."

So political extremists of all stripes offer utopian visions, which credulous idealists find remarkably attractive. But considering the disparity of the visions offered—the perfect market of laissez-faire, the perfect society of socialism, or the perfectly harmonious environment of eco-radicalism—it is not surprising that utopianism in the end only increases our social and intellectual rifts, steadily diminishing our chances of avoiding an ecological holocaust.

As I have argued throughout this work, social polarization is itself one of the central threats to the global environment. Most eco-radicals, however, disagree sharply. In fact, even moderate environmentalists often argue that green extremism serves a useful purpose by making other forms of environmentalism seem moderate in comparison, hence more palatable to a large segment of the population. But according to this logic, any vociferous, ideological minority ought to be easily able to pull public opinion in its own direction. This is, I strongly assert, a naive and dangerous belief. Extreme positions usually provoke fervent opposition among nonbelievers, not partial, lukewarm conversion. Do the aggressive and outrageously anti-environmental proposals of the "wise-use movement" convince environmentalists that George Bush is really on their side after all?

In conclusion, environmentalism's challenge must be more than to criticize society and imagine a blissful alternative. On the contrary, the movement must devise realistic plans and concrete strategies for avoiding ecological collapse and for reconstructing an ecologically sustainable economic order. To do so will entail working with, not against, society at large.

The best hope I see is through a new alliance of moderates from both the left and the right—a coalition in which moderate conservatives continue to insist on efficiency and prudence, and where liberals forward an agenda aimed at social progress and environmental protection, but in which both contingents are willing to compromise in the interests of a common nation and, ultimately, a common humanity. The environmental reforms necessary to ensure planetary survival will require the forging of such a broad-ranging political consensus. By thwarting its development, eco-radicalism undermines our best chance of salvaging the earth—offering instead only the peace of mind that comes from knowing that one's own ideology is ecologically and politically pure. It is time for the environmental movement to recognize such thinking for the fantasy that it is. We must first relinquish our hopes for utopia if we really wish to save the earth.

Promethean environmentalism is not simply a watered down, compromised form of the radical doctrine. Although its concrete proposals and its philosophical positions are consistently at odds with those of eco-radicalism (see the appendix), its ultimate purpose is in fact the same: to return the surface of the earth to *life*, to life in all its abundance, diversity, and evolutionary potential. Prometheans maintain, however, that for the foreseeable future we must *actively manage* the planet to ensure the survival of as much biological diversity as possible. No less is necessary if we are to begin atoning for our very real environmental sins—for our fall from grace that began at the end of the Pleistocene epoch.

Eco-radicalism tells us that we must dismantle our technological and economic system, and ultimately our entire civilization. Once we do so, the rifts between humanity and nature will purportedly heal automatically. I disagree. What I believe we must do is disengage humanity from nature by cleaving to, but carefully guiding, the path of technological progress. It is for the environmental community to decide which alternative offers the best hope for ecological salvation.

Appendix

■ A Tabular Comparison of Radical Environmentalism and Promethean
Environmentalism (Note: the following tables focus on the dominant,
Arcadian strain of radical environmentalism, so-called deep ecology)

Table A.1   General Orientations

Arcadian Environmentalism (Exemplified by the deep ecology of antihumanist anarchism)	Promethean Environmentalism
Utopian politics	The politics of the possible
Uncompromising standards	Compromise and conciliation embraced
Steady-state economic equilibrium	Economic growth based on increasing efficiency
Radical restructuring of society	Continual reform
Reimmersion of humankind within nature	Decoupling of humankind from nature
Philosophical idealism	Middle ground between idealism and materialism
Subversive postmodernism *or* pantheistic rationalism	Moderate postmodernism
Inverted orientalism (the West as the root of all evil)	Cosmopolitanism (denial of Western exceptionalism)
American democracy considered a sham	American democracy viewed as imperfect but real
Biocentric egalitarianism	Intrinsic value of all forms of life— without pan-species egalitarianism
(Normative) environmental determinism	Environmental possibilism

Table A.2   Anthropological and Ecological Positions

Arcadian Environmentalism	Promethean Environmentalism
Social Edenism (belief that all primal peoples enjoy perfect social harmony)	Good and evil seen as attributes of all human societies
Environmental Edenism (belief that primal peoples enjoy perfect environmental harmony)	Environmental destructiveness considered a potential attribute of all societies
Ecological equilibrium prevalent	Ecological flux prevalent
Holistic apprehension of nature	Both synthetic and reductive modes of inquiry considered vital
Pristine nature existing where undisturbed by modern humanity	Virtually all landscapes considered fundamentally anthropogenic
Human management of nature considered arrogant and destructive	Human management of nature considered necessary for the preservation of biodiversity
Participatory consciousness seen as a uniform characteristic of primal peoples; analytic consciousness a uniform attribute of modern peoples	Psychic universalism—tempered by cultural diversity
Primal peoples considered truly affluent	Tribal peoples viewed as often penurious
Human social organization comprehensible through ecological analogies	Human social organization requires social explanations

Table A.3   Interpretations of Scale in the Human Endeavor

Arcadian Environmentalism	Promethean Environmentalism
Small is beautiful/large is ugly	Small is sometimes ugly/large is sometimes beautiful
Anti-urbanism	Pro-urbanism
Low-density settlement advocated	High-density settlement advocated
Economic and political autarky	Economic and political integration
Bioregionalism	Political-ecological integration
"Think globally: act locally"	"Think and act at multiple spatial scales"
No middle ground between utopian decentralization and totalitarian centralization	Desirability of coordinated decentralization and flexible centralization
Communications technology enhances surveillance, hence oppression	Communications technology enhances freedom

Table A.3   *Continued*

Arcadian Environmentalism	Promethean Environmentalism
Present trends indicate increasing global cultural uniformitization	Present trends also indicate global cultural hybridization and persistence of diversity

Table A.4   Views on Technology

Arcadian Environmentalism	Promethean Environmentalism
Craft production advocated	Flexible automation advocated
Natural products created from once-living organisms advocated	Artificial products created from inert materials advocated
Communications by word of mouth and conventional print	Communications by advanced telecommunications
Energy from simple solar collectors and biomass	Energy from advanced solar collectors
Amish-style farming	Integrated pest management; greenhouse cultivation; ersatz foodstuffs; genetic engineering
Technological advance viewed as intrinsically destructive	Technological advance viewed as dangerous yet potentially salvaging
Science viewed as a force that alienates people from nature	Science viewed as necessary for environmental salvation
Eliminate toxic waste by eliminating modern industry	Eliminate toxic waste by improving technology and tightening regulations

Table A.5   Economic Positions

Arcadian Environmentalism	Promethean Environmentalism
Barter economy	Guided capitalism
Reduction of living standards in the First World	Improvement of living standards throughout the world
Constrained investments necessary	Enhanced investments necessary
Unemployment results from automation	Unemployment results from lagging productivity
Cooperation endorsed, competition denounced	Competitive cooperation endorsed
Capitalism viewed as intrinsically deadly to nature and society	Capitalism viewed as both potentially destructive and potentially salvaging

Table A.5 *Continued*

Arcadian Environmentalism	Promethean Environmentalism
Capitalism viewed as a unitary phenomenon that cannot be reformed	Capitalism viewed as a multifaceted phenomenon that is amenable to genuine reform
Capitalism viewed as *necessarily* preoccupied with the short term	Capitalism viewed as a system that works best when adopting a long-term perspective
Capitalism viewed as a system that will (soon) self-destruct	Capitalism viewed as remarkably adaptive
Capitalism viewed as *necessarily* tending toward ever greater centralization	Capitalism viewed as having both centrifugal and centripetal tendencies

Table A.6 Perspectives on Global Development

Arcadian Environmentalism	Promethean Environmentalism
First World wealth based solely on exploitation of the Third World	First World wealth based substantially on internal economic dynamics
Industrialization of the Third World to be sedulously avoided	Industrialization of the Third World necessary for social justice and environmental protection
Development must be based on low-tech, agrarian initiatives	Development must be based on a combination of low-tech and high-tech agrarian and industrial initiatives
Third World countries should strive for economic self-sufficiency	Third World countries should strive for integration within the world economy—but with favorable terms

Notes

■ 1   The Varieties of Radical Environmentalism

1   Not surprisingly, the eco-radical movement is marked by strong internal disagreements over ethical principles. Whereas outsiders often assume that a strong affinity exists between radical environmentalists and members of the animal rights movement, the two groups are uneasy allies at best. Animal rights advocates typically limit their concern to sentient beings (see Singer 1975), a restriction that most eco-radicals find highly prejudicial. The main body of the eco-radical movement bases its ethical concerns at the ecosystem level, stressing not individual entities at all but rather mutual relations. According to Nash (1989:5), this more encompassing vision reveals a progressive expansion of the ethical universe, a realm that begins with the self but that may ultimately include the entire universe. Such holistic moral reasoning easily leads to a mystical extension of consciousness to complex features of the natural world, such as rivers and mountains (e.g., Devall and Sessions 1985:112). Many environmental philosophers, however, reject such mysticism. Brennan (1988), for example, argues that while we should focus our ethical concerns on ecosystems rather than individuals, we would best avoid the pitfall of a quasi-spiritual systemic understanding that denies the reality of individual entities. In the most philosophically rigorous exposition of environmental ethics to date, Paul Taylor (1986) denies the extension of ethical principles to inanimate objects, and on the whole is quite skeptical of ecological holism, but he vigorously upholds the principle of biotic egalitarianism (believing that all forms of life are deserving of equal moral consideration). As a result he is implacably opposed to sport hunting and even fishing; activities supported by many eco-radicals as exemplifying natural processes. More recently, Stone (1988) has suggested that environmentalists abandon the quest for a unitary set of moral principles and instead embrace moral pluralism, a stance vigorously countered by Callicott (1990).

2   In his recent work *Confessions of an Eco-Warrior*, Foreman (1991) takes a more accommodating stance and seeks nonviolent approaches where possible.

3 In many European countries marxian political parties are both more powerful and more concerned with environmental issues than are their counterparts in the United States. This is especially true in Italy, where the Communist Party "has shown the most genuine concern for environmental issues, particularly in some of the communities and regions it governs" (Liberatore and Lewanski 1990:37).

■ 2 Primal Purity and Natural Balance

1 Guthrie argues that horses may have been able to readapt to the environment of the American West due to landscape changes wrought over the past 10,000 years by human beings (1984:289, 290), a thesis far-fetched but difficult to falsify.
2 Gimbutas's thesis has been recently challenged on archaeological and linguistic grounds, most notably by Renfrew (1987). According to Renfrew, Indo-European languages originated in Anatolia and spread in a generally peaceful process. And as Mallory (1989:258) cautions in his balanced account of Indo-European origins: "We are always wary of suggesting models of expansion that will be characterized as hordes of frenzied Aryans bursting out of the Russian steppe and slashing their way into the comparative grammars of historical linguists."

■ 3 A Question of Scale

1 This famous slogan was apparently popularized by an eco-religious sect in California called Creative Initiative which later transformed itself into the antiwar federation known as Beyond War (Gelber and Cook 1990:188).
2 Sale assiduously avoids the tainted term "hierarchy," preferring instead the locution "like Chinese boxes" (1985:56).
3 It is also apposite to note that the (direct) democratic party of Pericles in ancient Athens was also the main force behind the policy of imperial expansion and exploitation—a policy that eventually brought ruin to Athens.
4 Dave Foreman (1991:79), however, apparently does believe in respecting game laws.
5 Population density figures derived from estimated 1989 population figures found in the California Statistical Abstract (1989:14–18; table B-4), and from land area figures found in the California Statistical Abstract (1973, section A, p. 1; "Area, Geography and Climate"); areas of inland water were excluded from the calculations. The following counties were tabulated: over 2,000 persons per square mile, San Francisco; 1,000–2,000 persons per square mile, Alameda, Contra Costa, Sacramento, San Mateo, and Santa Clara; 200–1,000 persons per square mile, Marin, San Joaquin, Santa Cruz, Solano, Sonoma, and Stanislaus; 50–200 persons per square mile, Amador, Butte, El Dorado, Fresno, Kings, Merced, Monterey, Napa, Nevada, Placer, Sutter, Tulare, Yolo, and Yuba; 10–50 persons per square mile, Calaveras, Colusa, Del Norte, Glenn, Humboldt, Lake, Madera, Mariposa, Mendocino, Shasta, Tehama, and Toulumne; and under 10 persons per square mile, Alpine, Inyo, Lassen, Modoc, Mono, Plumas, Sierra, Siskiyou, and Trinity. Voting figures obtained from "Certified Statement of Vote, Nov. 6, 1990, General Election," compiled by March Fong Eu, secretary of state, and "Certified Statement of Vote, June 5, 1990, Primary Election," compiled by March Fong Eu, secretary of state.

Southern California counties are excluded by virtue of their geographical patterns: the immense county of San Bernardino, for example, has large areas of metropolitan suburbs, yet even larger areas of unpopulated desert.

6 The one saving grace of suburbs is that their residents tend to vote in favor of environmental issues, as is evident in tables 1 and 2.

7 We must also note that, with due respect to Reich, the transnationalization of capitalism may soon face insurmountable limits; Michael Porter (1990:19), in fact, argues that globalization will only make a given firm's home base ever more important.

■ 4 Technophobia and Its Discontents

1 The commercialization of continuous-fiber ceramic composites could, in fact, result in an energy savings of .7 quadrillion Btu by the year 2010 (Hirst 1991:32).

2 As cold-blooded animals, fish convert vegetable matter into flesh much more efficiently than do mammals and birds.

3 Biotechnology can be employed for environmental benefits in many areas other than agriculture or toxic waste decomposition. Biohydrometallurgy, for example, employs bacteria to remove sulfur from coal or mine wastes, and it can even be used to extract copper from ore bodies, obviating the need for environmentally destructive smelting. In the future, advocates of the process foresee the application of microorganism-based techniques to "*in situ* mining, which would leave the surrounding environment relatively undisturbed while removing the desired metals" (Debus 1990:55). For the present, biohydrometallurgy has already been responsible both for saving the American copper industry from financial collapse and for reducing significantly its environmental impact.

■ 5 The Capitalist Imperative

1 The prominence of marxist geography is clearly evident in lists of most-cited articles; see, for example, Whitehand (1990:21).

2 Virtually all that Seldon (1990:126) writes about Japan is that its successful firms often ignore governmental directives—a truth so partial as to be meaningless.

3 Several writers have argued that no German capitalists actually supported Hitler in his early rise to power (for example, Hall 1985:168), but scholarly opinion on this matter is still divided.

4 This "sports analogy," however, can be taken too far: unsuccessful athletic teams, for example, do not go bankrupt and disband. Despite both its intrinsically competitive nature and the astronomical salaries of its star players, major-league athletics is in some ways rather socialistic—just try to imagine a draft system in which unprofitable firms get the first choice on hot young MBAS!

5 One recent study does show, however, that plants with worker participation committees sometimes prove less efficient than those without them (see *Business Week*, April 1, 1991, p. 18).

6 It should be noted, however, that a recent MIT study indicates that gas taxes may not be as regressive as previously thought (see *Business Week*, April 8, 1991, p. 16).

7 Writers for *The Economist* fear that Japan and Germany may be beginning to emulate the Anglo-Saxon model of "punter-capitalism" that emphasizes stock

market gambling. "If the two pairs of countries keep moving in their present directions" they inform us, "it is only a matter of time before Japanese and German firms are systematically squandering their capital" ("Survey of Capitalism," May 5, 1990, p.17).

■ 6 Third World Development and Population

1 Whatever one thinks of Commoner's views on development, one must admit that his assertion that more people live in the Southern Hemisphere than the Northern Hemisphere betrays an appalling ignorance of world geography.
2 O'Riordan (1988:30, 48) intriguingly mixes both views by arguing that sustainability is "politically treacherous since it challenges the status quo," but that it "can be manipulated into tinkering adjustments to the status quo by established interests."
3 There are, however, a number of successful rural Third World environmental associations, most notably the *Chipko* forest-preservation movement of India, and the rubber-tapper organizations of the upper Amazon Basin.
4 Historically too, it is necessary to realize that even in a few colonized zones, participation in the world economy brought some benefits. Indeed, the socialist historian David Washbrook (1990) verges, at points, on describing India as Britain's partner in empire, and he shows clearly how British ideas of private property helped nurture an indigenous and progressive Indian capitalism.
5 The Latin American experience, however, would caution against such a heavy reliance on direct foreign investments. It is quite possible that Singapore has found such a strategy advantageous only because its exiguous size has precluded the development of a more nationalistic form of capitalism.
6 It must also be noted, however, that even in the absence of protection, export-led growth is never secure. The development of fully automated textile technology, for example, could result in the transfer of production back to the First World.
7 In some areas, however, land degradation may stem from population decline. When opportunities increase for wage migration, the labor formerly devoted to maintaining intensive, soil-conserving forms of agriculture may be no longer available, resulting in accelerated soil erosion.
8 This gender difference in reproductive preference may even be partially rooted in biology. Sociobiologists, whose reductionistic and deterministic reasoning must be approached with considerable caution, argue that since males can have a virtually unlimited number of offspring they often consider it a worthwhile gamble to squander genetic material in hopes that at least a few of their many children will make it to adulthood and thus pass on their genes—giving little regard to the amount of care that children require. Females, on the other hand, who intrinsically have a much more limited potential for reproduction, find greater genetic advantage in investing heavily in a small number of children.

■ Conclusion

1 As Said (1978:42) makes it clear, "the essence of orientalism is the ineradicable distinction between western superiority and oriental inferiority." The inverse orientalism of the eco-radicals might therefore be more properly labeled "occidentalism."

Bibliography

Adams, Richard. 1977. *Prehistoric Mesoamerica*. Boston: Little, Brown.

Adams, W. M. 1990. *Green Development: Environment and Sustainability in the Third World*. London: Routledge.

Agnew, John, and Corbridge, Stuart. 1989. "The New Geopolitics: The Dynamics of Geopolitical Disorder." In R. J. Johnston and P. J. Taylor, eds., *A World in Crisis? Geographical Perspectives*, pp. 266–88. Oxford: Basil Blackwell.

Albert, Bruce. 1989. "Yanomami 'Violence': Inclusive Fitness of Ethnographic Representation." *Current Anthropology* 30:637–40.

Alexander, Donald. 1990. "Bioregionalism: Science or Sensibility?" *Environmental Ethics* 12:161–72.

Allen, Kelly. 1991. "One of the Last of the Best." *Nature Conservancy* 41:17–23.

Anderson, Elain. 1984. "Who's Who in the Pleistocene: A Mammalian Bestiary." In P. Martin and R. Klein, eds., *Quaternary Extinctions: A Prehistoric Revolution*, pp. 40–89. Tucson: University of Arizona Press.

Anderson, Perry, and Leal, Donald. 1991. *Free Market Environmentalism*. San Francisco: Pacific Research Institute for Public Policy.

Barash, David. 1991. *The L Word: An Unapologetic, Thoroughly Biased Long-Overdue Explication and Celebration of Liberalism*. New York: William Morrow Company.

Bargatzky, Thomas. 1984. "Culture, Environment, and the Ills of Adaptationalism." *Current Anthropology* 25:399–415.

Barbour, Ian, ed. 1973. *Western Man and Environmental Ethics*. Reading, Mass.: Addison-Wesley.

Barr, Brenton. 1988. "Perspectives on Deforestation in the U.S.S.R." In J. Richards and R. Tucker, eds., *World Deforestation in the Twentieth Century*. Durham, N.C.: Duke University Press.

Bateson, Gregory. 1972. *Steps to an Ecology of Mind*. New York: Ballantine.

———. 1979. *Mind and Nature: A Necessary Unity*. New York: Dutton.

Bauer, P. T. 1981. *Equality, the Third World, and Economic Delusion*. Cambridge, Mass.: Harvard University Press.

Beckerman, Stephan. 1979. "The Abundance of Protein in Amazonia: A Reply to Gross." *American Anthropologist* 81:533–60.

———. 1983. "Does the Swidden Ape the Jungle?" *Human Ecology* 11:1–11.

Bell, Thomas. 1990. "Political Economy's Response to Positivism." *Geographical Review* 80:308–15.

Berger, Peter. 1991 (1986). *The Capitalist Revolution: Fifty Propositions about Prosperity, Equality, and Liberty*. New York: Basic Books.

Bergman, R. W. 1980. *Amazon Economics: The Simplicity of Shipibo Indian Wealth*. Dellplain Latin American Studies (vol. 6). Syracuse, N.Y.: Syracuse University, Department of Geography.

Bernal, Martin. 1987. *Black Athena: The Afroasiatic Roots of Classical Civilization. Volume One: The Fabrication of Ancient Greece 1785–1985*. New Brunswick, N.J.: Rutgers University Press.

Bernstein, Gregg. n.d. "Photovoltaics: An Analysis of the Technology, the Industry, and the Prospects for Large-Scale Electrical Generation through the Use of Thin Films." Unpublished research paper, Department of Geography, George Washington University, Washington, D.C.

Berreman, Gerald. 1991. "The Incredible 'Tasaday': Deconstructing the Myth of a Stone Age People." *Cultural Survival Quarterly* 15:3–44.

Berry, B. J. L. 1967. *The Geography of Market Centers and Retail Distribution*. Englewood Cliffs, N.J.: Prentice-Hall.

Berry, Wendell. 1977. *The Unsettling of America*. San Francisco: Sierra Club.

———. 1981. *The Gift of the Good Land: Further Essays Cultural and Agricultural*. San Francisco: North Point.

———. 1990. *What Are People For?* San Francisco: North Point.

Bevington, Rick, and Rosenfeld, Arthur. 1990. "Energy for Homes and Buildings." *Scientific American* 263:76–86.

Birket-Smith, Kaj. 1971. *Eskimos*. New York: Crown.

Blaikie, Piers. 1985. *The Political Economy of Soil Erosion in Developing Countries*. New York: Longman.

———. 1989. "The Use of Natural Resources in Developing and Developed Countries." In R. J. Johnston and P. J. Taylor, eds., *A World in Crisis? Geographical Perspectives*, pp. 125–50. Oxford: Basil Blackwell.

Blaikie, Piers, and Brookfield, Harold. 1987. *Land Degradation and Society*. London: Methuen.

Blainey, Geoffrey. 1975. *Triumph of the Nomads: A History of Ancient Australia*. Melbourne: Macmillan of Australia.

Blaut, James. 1989. Review of *Capitalist World Development*, by Stuart Corbridge. *The Professional Geographer* 41:102–3.

Blinder, Alan. 1987. *Hard Heads, Soft Hearts: Tough-Minded Economics for a Just Society*. Reading, Mass.: Addison-Wesley.

Blum, Jerome. 1987. *The End of the Old Order in Rural Europe*. Princeton, N.J.: Princeton University Press.

Boas, Franz. 1940. *Race, Language and Culture*. New York: Free Press.

Bonner, Raymond. 1987. *Waltzing with a Dictator: The Marcoses and the Making of American Policy*. Quezon City, Philippines: KEN.

Bookchin, Murray. 1972. *Post-Scarcity Anarchism*. Montreal: Black Rose Books.

————. 1986. *The Modern Crisis*. Philadelphia: New Society.

————. 1989. *Remaking Society*. Montreal: Black Rose Books.

Booth, Annie, and Jacobs, Harvey. 1990. "Ties that Bind: Native American Beliefs as a Foundation for Environmental Consciousness." *Environmental Ethics* 12:27–47.

Borrelli, Peter, ed. 1988. *Crossroads: Environmental Priorities for the Future*. Covelo, Calif.: Island.

Boserup, Ester. 1965. *The Conditions of Agricultural Growth*. Chicago: Aldine.

Botkin, Daniel. 1990. *Discordant Harmonies: A New Ecology for the Twenty-First Century*. New York: Oxford University Press.

Boulding, Kenneth. 1973. "The Shadow of the Stationary State." In M. Olson and H. Landsberg, eds., *The No Growth Society*, pp. 89–102. New York: W. W. Norton.

Bradford, Colin. 1990. "Policy Interventions and Markets: Development Strategy Typologies and Policy Option." In G. Gereffi and D. Wyman, eds., *Manufacturing Miracles: Paths of Industrialization in Latin America and East Asia*, pp. 32–54. Princeton, N.J.: Princeton University Press.

Bradley, P. N., and Carter, S. E. 1989. "Food Production and Distribution—and Hunger." In R. J. Johnston and P. J. Taylor, eds., *A World in Crisis? Geographical Perspectives*, pp. 101–24. Oxford: Basil Blackwell.

Bramwell, Anna. 1989. *Ecology in the 20th Century: A History*. New Haven, Conn.: Yale University Press.

Braudel, Fernand. 1981. *The Structures of Everyday Life: The Limits of the Possible (Volume One: Civilization and Capitalism, 15th–18th Century)*. New York: Harper and Row.

————. 1982. *The Wheels of Commerce: Volume Two, Civilization and Capitalism 15th–18th Century*. New York: Harper and Row.

————. 1984. *The Perspective of the World: Volume Three, Civilization and Capitalism 15th–18th Century*. New York: Harper and Row.

————. 1988. *The Identity of France: Volume One, History and Environment*. New York: Harper and Row.

————. 1990. *The Identity of France: Volume Two, People and Production*. New York: Harper Collins.

Brennan, Andrew. 1988. *Thinking about Nature*. Athens: University of Georgia Press.

Brightman, Robert. 1987. "Conservation and Resource Depletion: The Case of the Boreal Forest Algonquians." In B. McCay and J. Acheson, eds., *The Question of the Commons: The Culture and Ecology of Communal Resources*, pp. 121–41. Tucson: University of Arizona Press.

Bromberger, Norman, and Hughes, Kenneth. 1987. "Capitalism and Underdevelopment in South Africa." In J. Butler, R. Elphick, and D. Welsh, eds., *Democracy and Liberalism in South Africa: Its History and Prospect*, pp. 203–23. Middletown, Conn.: Wesleyan University Press.

Brown, Charles, Hamilton, James, and Medoff, James. 1990. *Employers Large and Small*. Cambridge, Mass.: Harvard University Press.

Bunge, W. 1989. "Epilogue: Our Planet Is Big Enough for Peace but too Small for War." In R. J. Johnston and P. J. Taylor, eds., *A World in Crisis? Geographical Perspectives*, pp. 355–57. Oxford: Basil Blackwell.

Burgelis, Sonja. n.d. "Environmental Pollution in Latvia." Unpublished research paper, Department of Geography, George Washington University, Washington, D.C.

Cahn, Robert, ed. 1985. *An Environmental Agenda for the Future*. Covelo, Calif.: Island.

Cain, Mead. 1981. "Risk and Insurance." *Population and Development Review* 7:435–74.

Caldwell, J. C. 1978. "A Theory of Fertility: From High Plateau to Destabilization." *Population and Development Review* 4:553–77.

Callicott, J. Baird. 1980. "Animal Liberation: A Triangular Affair." *Environmental Ethics* 2:311–38.

———. 1982a. "Hume's Is/Ought Dichotomy and the Relation of Ecology to Leopold's Land Ethic." *Environmental Ethics* 4:163–74.

———. 1982b. "Traditional American Indian and Western European Attitudes toward Nature: An Overview." *Environmental Ethics* 4:293–318.

———. 1986. "The Metaphysical Implications of Ecology." *Environmental Ethics* 8:301–16.

———. 1990. "The Case against Moral Pluralism." *Environmental Ethics* 12:99–124.

Candelina, Rowe. 1988. "A Comparison of Batak and Ata Subsistence Styles in Two Different Social and Physical Environments." In A. T. Rambo, K. L. Hutterer, and K. Gillogly, eds., *Ethnic Diversity and the Control of Natural Resources in Southeast Asia*, pp. 59–82. Michigan Papers on South and Southeast Asia, Center for South and Southeast Asian Studies (vol. 32). Ann Arbor: University of Michigan.

Cantor, Norman. 1988. *Twentieth-Century Culture: Modernism to Deconstruction*. New York and Berne: Peter Lang.

Carrier, James. 1987. "Marine Tenure and Conservation in Papua New Guinea: Problems in Interpretation." In B. McCay and J. Acheson, eds., *The Question of the Commons: The Culture and Ecology of Communal Resources*, pp. 142–70. Tucson: University of Arizona Press.

Catton, William. 1980. *Overshoot: The Ecological Basis of Revolutionary Change*. Urbana: University of Illinois Press.

Center for the Defense of Free Enterprise. 1990. *The Wise Use Memo*. Bellevue, Wash.

Chagnon, Napoleon. 1968. *Yanomamo: The Fierce People*. New York: Holt, Reinhart and Winston.

———. 1990. "On Yanomamo Violence: Reply to Albert." *Current Anthropology* 31:49–53.

Chagnon, Napoleon, and Hames, Raymond. 1979. "Protein Deficiency and Tribal Warfare in Amazonia: New Data." *Science* 203:910–13.

Chance, Norman. 1990. *The Inupiat*. Fort Worth: Holt, Rinehart and Winston.

Chandler, Alfred. 1990. *Scale and Scope: The Dynamics of Industrial Capitalism*. Cambridge, Mass.: Belknap Press of Harvard University Press.

"The Changing Face of Environmentalism in the Soviet Union." 1990. *Environment* 23 (2):2ff.

Cheney, Jim. 1987. "Eco-Feminism and Deep Ecology." *Environmental Ethics* 9:115–46.

———. 1989a. "Postmodern Environmental Ethics: Ethics as Bioregional Narrative." *Environmental Ethics* 11:117–34.

———. 1989b. "The Neo-Stoicism of Radical Environmentalism." *Environmental Ethics* 11:293–325.

Cheng, Tun-jen. 1990. "Political Regimes and Development Strategies: South Korea and Taiwan." In G. Gereffi and D. Wyman, eds., *Manufacturing Miracles: Paths of*

Industrialization in Latin America and East Asia, pp. 139–78. Princeton, N.J.: Princeton University Press.

Chuang, Chin. 1988. "Perspective of Industrial Waste Disposal in Taiwan." *Indian Journal of Environmental Protection* 8:401–8.

Clark, John. 1989. "Marx's Inorganic Body." *Environmental Ethics* 11:243–58.

Cohen, Mark. 1977. *The Food Crisis in Prehistory: Overpopulation and the Origins of Agriculture*. New Haven, Conn.: Yale University Press.

———. 1989. *Health and the Rise of Civilization*. New Haven, Conn.: Yale University Press.

Commoner, Barry. 1990. *Making Peace with the Planet*. New York: Pantheon.

Conklin, Harold. 1954. "An Ethnoecological Approach to Shifting Cultivation." *New York Academia of Sciences, Transactions* 17:133–42.

Connell, J. H. 1978. "Diversity in Tropical Rain Forests and Coral Reefs." *Science* 199:1302–9.

Conner, Daniel. 1987. "Is AIDS the Answer to an Environmentalist's Prayer?" *Earth First!* December 22, 1987, pp. 14–16.

Conway, William. 1988. "Can Technology Aid Species Preservation?" In E. O. Wilson and F. Peter, eds., *Biodiversity*, pp. 263–68. Washington, D.C.: National Academy.

Conzen, Michael. 1987. "The Progress of American Urbanism, 1860–1930." In R. Mitchell and P. Groves, eds., *North America: The Historical Geography of a Changing Continent*, pp. 347–72. Totowa, N.J.: Rowman and Littlefield.

Corbridge, Stuart. 1986. *Capitalist World Development: A Critique of Radical Development Geography*. London: Macmillan.

———. 1988. "Deconstructing Determinism: A Reply to Michael Watts." *Antipode* 20:239–59.

———. 1989. "Marxism, Post-Marxism, and the Geography of Development." In R. Peet and N. Thrift, eds., *New Models in Geography: The Political-Economy Perspective*, pp. 224–56. London: Unwin Hyman.

———. 1990. "Post-Marxism and Development Studies: Beyond the Impasse." *World Development* 18:623–39.

Crapanzano, Vincent. 1986. "Hermes' Dilemma: The Masking of Subversion in Ethnographic Description." In J. Clifford and G. Marcus, eds., *Writing Culture: The Poetics and Politics of Ethnography*, pp. 51–76. Berkeley: University of California Press.

Critchfield, Richard. 1983. *Villages*. Garden City, N.J.: Anchor.

Cronon, William. 1990. "Modes of Prophesy and Production: Placing Nature in History." *Journal of American History* 76:1,122–31.

———. 1991. *Nature's Metropolis: Chicago and the Great West*. New York: W. W. Norton.

Crosby, Alfred. 1986. *Ecological Imperialism: The Biological Expansion of Europe, 900–1900*. Cambridge: Cambridge University Press.

Cuzan, Alfred. 1983. "Appropriators versus Expropriators: The Political Economy of Water in the West." In T. Anderson, ed., *Water Rights: Scarce Resource Allocation, Bureaucracy, and the Environment*, pp. 13–44. San Francisco: Pacific Institute for Public Policy Research.

Daly, Herman. 1977. *Steady-State Economics: The Economics of Biophysical Equilibrium*. San Francisco: W. H. Freeman.

Daly, Herman, and Cobb, John. 1989. *For the Common Good: Redirecting the Econ-*

omy toward Community, the Environment, and a Sustainable Future. Boston: Beacon.

Dasmann, Raymond. 1988. "Toward a Biosphere Consciousness." In D. Worster, ed., *The Ends of the Earth: Perspectives on Modern Environmental History*, pp. 277–88. Cambridge: Cambridge University Press.

Davis, Ged. 1990 "Energy for Planet Earth." *Scientific American* 263:54–63.

Davis, Charles, and Lester, James. 1987. "Decentralizing Federal Environmental Policy: A Research Note." *Western Political Quarterly* 40:555–65.

Debus, Keith. 1990. "Mining with Microbes." *Technology Review*, August–September 1990, 51–57.

Degler, Carl. 1991. *In Search of Human Nature.* New York: Oxford University Press.

Deleage, Jean-Paul. 1989. "Eco-Marxist Critique of Political Economy." *Capitalism, Nature, Socialism*, 3:15–31.

Dennett, Glenn, and Connell, John. 1988. "Acculturation and Health in the Highlands of Papua New Guinea." *Current Anthropology* 29:273–99.

Dertouzos, Michael, Lester, Richard, and Solow, Robert. 1989. *Made in America: Regaining the Productive Edge.* Cambridge, Mass.: MIT Press.

De Soto, Hernando. 1989. *The Other Path: The Invisible Revolution in the Third World.* New York: Harper and Row.

Devall, Bill. 1988. *Simple in Means, Rich in Ends: Practicing Deep Ecology.* Salt Lake City: Peregrine Smith.

Devall, Bill, and Sessions, George. 1985. *Deep Ecology.* Salt Lake City: Peregrine Smith.

Dionne, E. J. 1991. *Why Americans Hate Politics.* New York: Simon and Schuster.

Dobson, Andrew. 1990. *Green Political Thought.* London: Unwin Hyman.

Dore, Ronald. 1990. "Reflections on Culture and Social Change." In G. Gereffi and D. Wyman, eds., *Manufacturing Miracles: Paths of Industrialization in Latin America and East Asia*, pp. 353–67. Princeton, N.J.: Princeton University Press.

Doubiago, Sharon. 1989. "Mama Coyote Talks to the Boys." In J. Plant, ed., *Healing the Wounds: The Promise of Ecofeminism*, pp. 40–44. Philadelphia: New Society.

Douglas, Mary. 1966. *Purity and Danger: An Analysis of Concepts of Pollution and Taboo.* London: Routledge and Kegan Paul.

Doyle, Michael. 1986. *Empires.* Ithaca, N.Y.: Cornell University Press.

Drexler, K. Eric. 1986. *Engines of Creation: The Coming Era of Nanotechnology.* New York: Doubleday.

Drexler, K. Eric, and Peterson, Chris. 1991. *Unbounding the Future: The Nanotechnology Revolution.* New York: William Morrow.

Dryzek, John. 1987. *Rational Ecology: Environment and Political Economy.* Oxford: Basil Blackwell.

D'Souza, Dinesh. 1991. *Illiberal Education: The Politics of Race and Sex on Campus.* New York: Free Press.

Duncan, James. 1980. "The Superorganic in American Cultural Geography." *Annals of the Association of American Geography* 70:181–98.

Eccleston, Bernard. 1989. *State and Society in Post-War Japan.* Cambridge: Polity.

Eckersley, Robyn. 1989. "Divining Evolution: The Ecological Ethics of Murray Bookchin." *Environmental Ethics* 11:99–116.

Ehrenfeld, David. 1978. *The Arrogance of Humanism.* New York: Oxford University Press.

Ehrlich, Paul. 1988. "The Loss of Diversity: Causes and Consequences." In E. O. Wilson and F. Peter, eds., *Biodiversity*, pp. 21–27. Washington, D.C.: National Academy.

———. 1989. "The Limits to Substitution: Meta-Resource Depletion and a New Economic-Ecological Paradigm." *Ecological Economics* 1:9–16.

Ehrlich, Paul, and Ehrlich, Anne. 1990. *The Population Explosion*. New York: Simon and Schuster.

Elkington, John, and Shopley, Jonathan. 1988. *The Shrinking Planet: U.S. Information Technology and Sustainable Development*. Washington, D.C.: World Resources Institute.

Ellen, Roy. 1982. *Environment, Subsistence, and System: The Ecology of Small-Scale Social Formations*. Cambridge: Cambridge University Press.

Ellison, Christopher, and Gereffi, Gary. 1990. "Explaining Strategies and Patterns of Industrial Development." In G. Gereffi and D. Wyman, eds., *Manufacturing Miracles: Paths of Industrialization in Latin America and East Asia*, pp. 368–403. Princeton, N.J.: Princeton University Press.

Ellul, Jacques. 1964. *The Technological Society*. New York: Alfred A. Knopf.

Environmental Ethics. 1989. "The Gospel of Chief Seattle is a Hoax." *Environmental Ethics* 11:195–96.

Faber, Daniel, and O'Connor, James. 1989. "Rejoinders" in "Discussion: The Struggle for Nature." *Capitalism, Nature, Socialism* 1:174–78.

Fajnzylber, Fernando. 1990. "The United States and Japan as Models of Industrialization." In G. Gereffi and D. Wyman, eds., *Manufacturing Miracles: Paths of Industrialization in Latin America and East Asia*, pp. 323–52. Princeton, N.J.: Princeton University Press.

Fegan, Brian. 1982. "The Social History of a Central Luzon Barrio." In A. W. McCoy and E. C. de Jesus, eds., *Philippine Social History: Global Trade and Local Transformations*, pp. 91–130. Honolulu: University of Hawaii Press.

Ferguson, Denzel, and Ferguson, Nancy. 1983. *Sacred Cows at the Public Trough*. Bend, Ore: Maverick.

Ficket, Arnold, Gellings, Clark, and Lovins, Amory. 1990. "Efficient Use of Electricity." *Scientific American* 263:64–74.

Fischoff, Baruch. 1991. "Report from Poland: Science and Politics in the Midst of Environmental Disaster." *Environment* 33 (2):12ff.

Fitzsimmons, Margaret. 1989. "The Matter of Nature." *Antipode* 21:106–20.

Folke, Steen, and Sayer, Andrew. 1991. "What's Left to Do?: Two Views from Europe." *Antipode* 23:240–48.

Forbes, Dean, and Thrift, Nigel, eds. 1981. *The Socialist Third World: Urban Planning and Territorial Development*. Oxford: Basil Blackwell.

Foreman, Dave. 1991. *Confessions of an Eco-Warrior*. New York: Harmony.

Fox, Richard. 1969. "Professional Primitives: Hunters and Gatherers in Nuclear South Asia." *Man in India* 48:139–60.

Fox, Warwick. 1989. "The Deep Ecology-Feminism Debate and Its Parallels." *Environmental Ethics* 11:5–25.

———. 1990. *Toward a Transpersonal Ecology: Developing New Foundations for Environmentalism*. Boston: Shambhala.

Frank, A. G. 1969. *Capitalism and Underdevelopment in Latin America*. New York: Monthly Review.

Freeman, Harry. 1990. *Hazardous Waste Minimization*. New York: McGraw Hill.

Frenkel, Stephen. n.d. "Bioregionalism, Geography and Environmental Determinism." Unpublished research paper, Department of Geography, Syracuse University, N.Y.

Friedman, J. 1974. "Marxism, Structuralism and Vulgar Materialism." *Man* 9:444–69.

Fürer-Haimendorf, Christoph von. 1982. *Tribes of India: The Struggle for Survival*. Berkeley: University of California Press.

Gabriel, Trip. 1990. "If a Tree Falls in the Forest, They Hear It." *New York Times Magazine*, November 4, 1990, p. 34ff.

Gasser, Charles, and Fraley, Robert. 1989. "Applications of Genetic Engineering to Crop Improvement." *Science* 210:1293–99.

Geertz, Clifford. 1963. *Agricultural Involution: The Process of Ecological Change in Indonesia*. Berkeley: University of California Press.

Gelber, Steven, and Cook, Martin. 1990. *Saving the Earth: The History of a Middle-Class Millennarian Movement*. Berkeley: University of California Press.

Gereffi, Gary. 1990. "Paths of Industrialization: An Overview." In G. Gereffi and D. Wyman, eds., *Manufacturing Miracles: Paths of Industrialization in Latin America and East Asia*, pp. 3–31. Princeton, N.J.: Princeton University Press.

Gereffi, Gary, and Wyman, D., eds. 1990. *Manufacturing Miracles: Paths of Industrialization in Latin America and East Asia*. Princeton, N.J.: Princeton University Press.

Gernet, Jacques. 1968. *Ancient China: From the Beginnings to Empire*. Berkeley: University of California Press.

Ghatak, S. 1988. "Toward a Second Green Revolution in the Tropics: From Chemicals to New Biological Techniques for Sustained Economic Development." In R. K. Turner, ed., *Sustainable Environmental Management: Principles and Practice*, pp. 145–69. London: Bellhaven.

Gibson, Thomas. 1986. *Sacrifice and Sharing in the Philippine Highlands*. London: Athlone.

Gillis, Malcolm, and Repetto, Robert. 1988. "Conclusion: Findings and Policy Implications." In R. Repetto and M. Gillis, eds., *Public Policies and the Misuse of Forest Resources*, pp. 385–410. Cambridge: Cambridge University Press.

Gimpel, Jean. 1976. *The Medieval Machine: The Industrial Revolution in the Middle Ages*. New York: Penguin.

Gimbutas, Marija. 1980. "The Kurgan Wave #2 (c. 3400–3200 B.C.) into Europe and the Following Transformation of Culture." *Journal of Indo-European Studies* 8:273–316.

Glendinning, Chellis. 1990a. *When Technology Wounds: The Human Consequences of Progress*. New York: William Morrow.

———. 1990b. "Notes toward a Neo-Luddite Manifesto." *Utne Reader* March–April 1990, pp. 50–53.

Goldfarb, Jeffrey. 1991. *The Cynical Society: The Culture of Politics and the Politics of Culture in American Life*. Chicago: University of Chicago Press.

Goodwin, Grenville. 1969. *The Social Organization of the Western Apache*. Tucson: University of Arizona Press.

Goucher, Candice. 1988. "The Impact of German Colonial Rule on the Forests of Togo." In J. Richards and R. Tucker, eds., *World Deforestation in the Twentieth Century*, pp. 56–69. Durham, N.C.: Duke University Press.

Goudie, Andrew. 1981. *The Human Impact: Man's Role in Environmental Change.* Cambridge, Mass.: MIT Press.

Graham, Edward. 1947. *Land and Wildlife.* New York: Oxford University Press.

Graham, Russell, and Lundelius, Ernest. 1984. "Coevolutionary Disequilibrium and Pleistocene Extinctions." In P. Martin and R. Klein, eds., *Quaternary Extinctions: A Prehistoric Revolution*, pp. 223–49. Tucson: University of Arizona Press.

Greenburg, M. and Amer, S. 1989. "Self-Interest and Direct Legislation: Public Support of a Hazardous Waste Bond Issue in New Jersey." *Political Geography Quarterly* 8:67–78.

Gross, D. 1975. "Protein Capture and Cultural Development in the Amazon." *American Anthropologist* 77:526–49.

Grossman, Lawrence. 1984. *Peasants, Subsistence Ecology, and Development in the Highlands of New Guinea.* Princeton, N.J.: Princeton University Press.

Grousset, Rene. 1970. *The Empire of the Steppes: A History of Central Asia.* New Brunswick, N.J.: Rutgers University Press.

Guha, Ramachandra. 1989. "Radical American Environmentalism and Wilderness Preservation: A Third World Critique." *Environmental Ethics* 11:71–83.

Guiliday, John. 1984. "Pleistocene Extinctions and Environmental Change: Case Study of the Appalachians." In P. Martin and R. Klein, eds., *Quaternary Extinctions: A Prehistoric Revolution*, pp. 250–58. Tucson: University of Arizona Press.

Guthrie, R. Dale. 1984. "Mosaics, Allelochemics, and Nutrients: An Ecological Theory of Late Pleistocene Megafaunal Extinctions." In P. Martin and R. Klein, eds., *Quaternary Extinctions: A Prehistoric Revolution*, pp. 259–98. Tucson: University of Arizona Press.

Haggard, Stephan. 1990. *Pathways from the Periphery: The Politics of Growth in the Newly Industrializing Countries.* Ithaca, N.Y.: Cornell University Press.

Hall, John. 1985. *Powers and Liberties: The Causes and Consequences of the Rise of the West.* Berkeley: University of California Press.

Hamerow, Theodore. 1990. *From the Finland Station: The Graying of the Revolution in the Twentieth Century.* New York: Basic Books.

Hames, Raymond. 1987. "Game Conservation or Efficient Hunting?" In B. McCay and J. Acheson, eds., *The Question of the Commons: The Culture and Ecology of Communal Resources*, pp. 92–107. Tucson: University of Arizona Press.

Hanley, Susan, and Yamamura, Kozo. 1977. *Economic and Demographic Change in Preindustrial Japan, 1600–1868.* Princeton, N.J.: Princeton University Press.

Hardin, Garrett. 1968. "The Tragedy of the Commons." *Science* 162:1,243–48.

———. 1977. *The Limits to Altruism: An Ecologist's View of Survival.* Bloomington: Indiana University Press.

Hardoy, Jorge, and Satterthwaite, David. 1985. "Third World Cities and the Environment of Poverty." In R. Repetto, ed., *The Global Possible: Resources, Development and the New Century*, pp. 171–210. New Haven, Conn.: Yale University Press.

Harris, Marvin. 1974. *Cows, Pigs, Wars, and Witches: The Riddles of Culture.* New York: Vintage.

Hartshorne, Richard. 1939. *The Nature of Geography.* Lancaster, Penn.: Association of American Geographers.

Harvey, David. 1982. *The Limits to Capital.* Chicago: University of Chicago Press.

———. 1989. *The Condition of Postmodernity: An Inquiry into the Origins of Cultural Change.* Oxford: Basil Blackwell.

———. 1990. "Between Space and Time: Reflections on the Geographical Imagination." *Annals of the Association of American Geographers* 80:418–34.

Hawes, Gary. 1987. *The Philippine State and the Marcos Regime: The Politics of Export*. Ithaca, N.Y.: Cornell University Press.

Headland, Thomas, and Reid, Lawrence. 1989. "Hunter-Gatherers and Their Neighbors from Prehistory to the Present." *Current Anthropology* 30:43–66.

Hecht, S., and Cockburn, A. 1989. *The Fate of the Forest: Developers, Destroyers, and Defenders of the Amazon*. London: Verso.

Heilbroner, Robert. 1980. *An Inquiry into the Human Prospect (Updated and Reconsidered for the 1980s)*. New York: W. W. Norton.

Heiser, Charles. 1981. *Seed to Civilization: The Story of Food*. San Francisco: W. H. Freeman.

Henning, Daniel, and Mangun, William. 1989. *Managing the Environmental Crisis: Incorporating Competing Values in Natural Resource Administration*. Durham, N.C.: Duke University Press.

Higgins, I. J., ed. 1985. *Biotechnology: Principles and Applications*. Oxford: Blackwell Scientific.

Hirst, Eric. 1991. "Boosting U.S. Energy Efficiency through Federal Action." *Environment* 33 (2):6ff.

Hobsbawm, Eric. 1975. *The Age of Capital 1848–1875*. New York: New American Library.

Holmes, John. 1986. "The Organizational and Locational Structure of Production Subcontracting." In A. Scott and M. Storper, eds., *Production, Work, Territory: The Geographical Anatomy of Industrial Capitalism*, pp. 80–106. Boston: Allen and Unwin.

Hugill, Peter. 1988. "Structural Changes in the Core Regions of the World Economy, 1830–1945." *Journal of Historical Geography* 14:11–127.

Hungspreugs, Manuwadi. 1988. "Heavy Metals and Other Non-Oil Pollutants in Southeast Asia." *Ambio* 17:178–82.

Huntington, Ellsworth. 1915. *Civilization and Climate*. New Haven, Conn.: Yale University Press.

Hurst, Philip. 1990. *Rainforest Politics: Ecological Destruction in South-East Asia*. London: Zed.

Jackson, Kenneth. 1985. *Crabgrass Frontier: The Suburbanization of the United States*. New York: Oxford University Press.

Jackson, Wes, and Bender, Marty. 1984. "Investigations into Perennial Polyculture." In W. Jackson, W. Berry, and B. Coleman, eds., *Meeting the Expectations of the Land*, pp. 183–94. San Francisco: North Point.

Janzen, Daniel. 1988. "Tropical Dry Forests: The Most Endangered Major Tropical Ecosystem." In E. O. Wilson and F. Peter, eds., *Biodiversity*, pp. 130–37. Washington, D.C.: National Academy.

Jaynal, N.D. 1985. "Destruction of Water Resources—The Most Critical Ecological Crisis of East Asia." *Ambio* 14:95–98.

Jencks, Charles. 1987. *Post-Modernism: The New Classicism in Art and Architecture*. New York: Rizzoli.

Jochim, Michael. 1981. *Strategies for Survival: Cultural Behavior in an Ecological Context*. New York: Academic.

Johannes, R. E. 1981. *Words of the Lagoon: Fishing and Marine Lore in the Palau District of Micronesia*. Berkeley: University of California Press.

Johnson, Allen. 1982. "Reductionism in Cultural Ecology: The Amazon Case." *Current Anthropology* 23:413–28.

Johnson, Allen, and Earle, Timothy. 1987. *The Evolution of Human Societies: From Foraging Group to Agrarian State*. Stanford: Stanford University Press.

Johnson, Chalmers. 1982. *MITI and the Japanese Miracle: The Growth of Industrial Policy 1925–1975*. Stanford: Stanford University Press.

Johnston, R. J. 1989. "The Individual and the World-Economy." In R. J. Johnston and P. J. Taylor, eds., *A World in Crisis? Geographical Perspectives*, pp. 200–228. Oxford: Basil Blackwell.

Johnston, James, and Robinson, Susan. 1984. *Genetic Engineering and New Pollution Control Technologies*. Park Ridge, N.J.: Noyes.

Jones, Eric L. 1988. *Growth Recurring: Economic Change in World History*. Oxford: Clarendon.

Jung, Hwa Yol. 1983. "Marxism, Ecology, and Technology." *Environmental Ethics* 5:169–71.

Kassiola, Joel Jay. 1990. *The Death of Industrial Civilization: The Limits of Economic Growth and the Repoliticization of Advanced Industrial Society*. Albany: State University of New York Press.

Keesing, Roger. 1987. "*Ta'a Geni*: Women's Perspectives on Kwaio Society." In M. Strathern, ed., *Dealing with Inequality: Analyzing Gender Relations in Melanesia and Beyond*, pp. 33–62. Cambridge: Cambridge University Press.

Kheel, Marti. 1985. "The Liberation of Nature: A Circular Affair." *Environmental Ethics* 7:135–49.

Klitgaard, Robert. 1991. *Adjusting to Reality: Beyond "State versus Market" in Economic Development*. San Francisco: ICS.

Knauft, Bruce. 1987. "Reconsidering Violence in Simple Human Societies." *Current Anthropology* 28:457–500.

Knight, David. 1983. "The Dilemma of Nations in a Rigid State-Structured World." In N. Kliot and S. Waterman, eds., *Pluralism and Political Geography: People, Territory, and State*, pp. 114–37. London: Croom Helm.

Knox, Paul, and Agnew, John. 1989. *The Geography of the World Economy*. London: Edward Arnold.

Koestler, Arthur. 1978. *Janus: A Summing up*. New York: Vintage.

Kokszka, Leopold, and Flood, Jared. 1989. *Environmental Management Handbook: Toxic Chemical Materials and Wastes*. New York: Marcel Dekker.

Konner, Melvin, and Shostak, Marjorie. 1986. "Ethnographic Romanticism and the Idea of Human Nature." In M. Biesele, ed., *The Past and Future of !Kung Ethnography*, pp. 69–76. Hamburg: Helmut Buske.

Koppes, Clayton. 1988. "Efficiency, Equity, Esthetics: Shifting Themes in American Conservation." In D. Worster, ed., *The Ends of the Earth: Perspectives on Modern Environmental History*, pp. 230–51. Cambridge: Cambridge University Press.

Kopytoff, Igor. 1987. *The African Frontier: The Reproduction of Traditional African Societies*. Bloomington: Indiana University Press.

Kotlyakov, V. M. 1991. "The Aral Sea Basin: A Critical Environmental Zone." *Environment* 33 (1):4ff.

Kriedte, Peter. 1983. *Peasants, Landlords, and Merchant Capitalists: Europe and the World Economy, 1500–1800.* Cambridge: Cambridge University Press.

Krugman, Paul. 1990. *The Age of Diminished Expectations: U.S. Economic Policy in the 1990s.* Cambridge, Mass.: MIT Press.

Kuttner, Robert. 1991. *The End of Laissez-Faire: National Purpose and the Global Economy after the Cold War.* New York: Alfred A. Knopf.

Laarman, Jan. 1988. "Export of Tropical Hardwoods in the Twentieth Century." In J. Richards and R. Tucker, eds., *World Deforestation in the Twentieth Century*, pp. 147–63. Durham, N.C.: Duke University Press.

Langer, Elinor. 1990. "The American Neo-Nazi Movement Today." *The Nation*, July 16–23, 1990, pp. 82–107.

Lee, Donald. 1980. "On the Marxian View of the Relationship between Man and Nature." *Environmental Ethics* 2:3–46.

———. 1982. "Toward a Marxian Ecological Ethics: A Response to Two Critics." *Environmental Ethics* 4:339–42.

Lee, Richard. 1979. *The !Kung San: Men, Women and Work in a Foraging Society.* Cambridge: Cambridge University Press.

Lee, R., and DeVore, I., eds. 1968. *Man the Hunter.* Chicago: Aldine.

Lehman, David. 1991. *Signs of the Times: Deconstruction and the Fall of Paul de Man.* New York: Poseidon.

Leidner, Jacob. 1981. *Plastics Waste.* New York: Marcel Dekker.

Leonard, H. Jeffrey. 1988. *Pollution and the Struggle for the World Product: Multinational Corporations, Environment, and International Comparative Advantage.* Cambridge: Cambridge University Press.

Leopold, Aldo. 1949 [1966]. *A Sand County Almanac.* New York: Ballantine.

Levi-Strauss, Claude. 1966. *The Savage Mind.* Chicago: University of Chicago Press.

Lewis, Archibald. 1988. *Nomads and Crusaders* A.D. 1000–1368. Bloomington: Indiana University Press.

Lewis, Bernard. 1990. *Race and Slavery in the Middle East: An Historical Inquiry.* Oxford: Oxford University Press.

Lewis, H. T. 1982. *Patterns of Indian Burning in California: Ecology and Ethnohistory.* Ramona, Calif.: Balena.

Lewis, Martin. 1989. "Commercialization and Community Life: The Geography of Market Exchange in a Small-Scale Philippine Society." *Annals of the Association of American Geographers* 79:390–410.

———. 1991. "Elusive Societies: A Regional-Cartographical Approach to the Study of Human Relatedness." *Annals of the Association of American Geographers*.

———. 1992. *Wagering the Land: Ritual, Capital, and Environmental Degradation in the Cordillera of Northern Luzon, 1900–1986.* Berkeley: University of California Press.

Libecap, Gary. 1981. *Locking up the Range: Federal Land Controls and Grazing.* Cambridge, Mass.: Ballinger.

Liberatore, Angela, and Lewanski, Rudolf. 1990. "The Evolution of Italian Environmental Policy." *Environment* 32 (5):10ff.

Lightfoot-Klein, Hanny. 1990. *Prisoners of Ritual: An Odyssey into Female Genital Circumcision in Africa.* Binghamton, N.Y.: Haworth.

Lipietz, Alain. 1987. *Mirages and Miracles: The Crises of Global Fordism.* London: Verso.

Lipton, Merle. 1985. *Capitalism and Apartheid*. Aldershot, England: Gower.

Lodge, David. 1984. *Small World*. New York: Warner.

Lowry, William. 1992. *The Dimersions of Federalism: State Governments and Pollution Control Policies*. Durham, N.C.: Duke University Press.

McCay, B., and Acheson, J., eds. 1987. *The Question of the Commons: The Culture and Ecology of Communal Resources*. Tuscon: University of Arizona Press.

McDonald, Jerry. 1984. "The Reordered North American Selection Regime and Late Quaternary Megafaunal Extinctions." In P. Martin and R. Klein, eds., *Quaternary Extinctions: A Prehistoric Revolution*, pp. 404–39. Tucson: University of Arizona Press.

McEvedy, Colin. 1961. *The Penguin Atlas of Medieval History*. New York: Penguin.

McEvedy, Colin, and Jones, Richard. 1978. *Atlas of World Population History*. New York: Facts on File.

McKean, Roland. 1973. "Growth vs. No Growth: An Evaluation." In M. Olson and H. Landsberg, eds., *The No Growth Society*, pp. 207–28. New York: W. W. Norton.

McKibben, Bill. 1989. *The End of Nature*. New York: Random House.

McLennan, Marshall. 1980. *The Central Luzon Plain: Land and Society on the Inland Frontier*. Quezon City, Philippines: Alemar-Phoenix.

MacNeill, Jim, Winsemius, Pieter, and Yakushiji, Taizo. 1991. *Beyond Interdependence: The Meshing of the World's Economy and the Earth's Ecology*. Oxford: Oxford University Press.

Maddox, John. 1972. *The Doomsday Syndrome*. New York: McGraw-Hill.

Mallory, J. P. 1989. *In Search of the Indo-Europeans*. London: Thames and Hudson.

Manes, Christopher. 1990. *Green Rage: Radical Environmentalism and the Unmaking of Civilization*. Boston: Little, Brown.

Mann, Michael. 1986. *The Sources of Social Power: Volume One, a History of Power from the Beginning to* A.D. 1760. Cambridge: Cambridge University Press.

Marietta, Don. 1988. "Environmental Holism and Individuals." *Environmental Ethics* 10:251–58.

Martin, Calvin. 1978. *Keepers of the Game: Indian-Animal Relationships and the Fur Trade*. Berkeley: University of California Press.

Martin, Paul. 1984. "Prehistoric Overkill: The Global Model." In P. Martin and R. Klein, eds., *Quaternary Extinctions: A Prehistoric Revolution*, pp. 354–403. Tucson: University of Arizona Press.

Martin, Paul, and Klein, Richard, eds. 1984. *Quaternary Extinctions: A Prehistoric Revolution*. Tucson: University of Arizona Press.

Martin, Paul, and Wright, H. E., eds. 1967. *Pleistocene Extinctions: The Search for a Cause*. New Haven, Conn.: Yale University Press.

Matossian, Mary. 1989. *Poisons of the Past: Molds, Epidemics, and History*. New Haven, Conn.: Yale University Press.

Meadows, Donella, Meadows, Dennis, Randers, Jorgen, and Behrens, William. 1972. *The Limits to Growth*. New York: Universe.

Merchant, Carolyn. 1989. *Ecological Revolutions: Nature, Gender, and Science in New England*. Chapel Hill: University of North Carolina Press.

Milbrath, Lester. 1989. *Envisioning a Sustainable Society: Learning Our Way Out*. Albany: State University of New York Press.

Miles, Gary. 1990. "Roman and Modern Imperialism: A Reassessment." *Comparative Studies in Society and History* 32:629–59.

Minshull, R. 1967. *Regional Geography: Theory and Practice*. London: Hutchinson University Library.

Mintz, Sidney. 1985. *Sweetness and Power: The Place of Sugar in Modern History*. New York: Penguin.

Mishan, E. J. 1973. "Ills, Bads, and Disamenities: The Wages of Growth." In M. Olson and H. Landsberg, eds., *The No Growth Society*, pp. 63–88. New York, W. W. Norton.

Mokyr, Joel. 1990. *The Lever of Riches: Technological Change and Economic Progress*. New York: Oxford University Press.

Moore, John. 1987. *The Cheyenne Nation*. Lincoln: University of Nebraska Press.

Mosher, Steven. 1983. *Broken Earth: The Rural Chinese*. New York: Free Press.

Moss, Mitchel. 1988. "Telecommunications: Shaping the Future." In G. Sternlieb and J. Hughes, eds., *America's New Market Geography: Nation, Region, and Metropolis*. New Brunswick, N.J.: Center for Urban Policy Research.

Moulder, Frances. 1977. *Japan, China, and the Modern World Economy*. Cambridge: Cambridge University Press.

Mowery, David, and Rosenberg, Nathan. 1989. *Technology and the Pursuit of Economic Growth*. Cambridge: Cambridge University Press.

Mumford, Lewis. 1966. *Technics and Human Development: Volume One, The Myth of the Machine*. New York: Harcourt Brace Jovanovich.

Murphy, Alexander. 1991. "The Emerging Europe of the 1990s." *Geographical Review* 81:1–17.

Murray, Martin. 1988. "The Triumph of Marxist Approaches in South African Social and Labour History." *Journal of Asian and African Studies* 23:78–101.

Myers, Raymon, and Peatie, Mark. eds. 1984. *The Japanese Colonial Empire, 1895–1945*. Princeton, N.J.: Princeton University Press.

Naess, Arne. 1989. *Ecology, Community and Lifestyle: Outline of an Ecosophy*. Translated and edited by David Rothenberg. Cambridge: Cambridge University Press.

Nance, John. 1975. *The Gentle Tasaday*. New York: Harcourt Brace Jovanovich.

Nash, Roderick. 1989. *The Rights of Nature: A History of Environmental Ethics*. Madison: University of Wisconsin Press.

National Research Council. 1989. *Field Testing Genetically Modified Organisms: Framework for Decisions*. Washington, D.C.: National Academy.

Nelson, K. 1986. "Labor Demand, Labor Supply and the Suburbanization of Low-Wage Office Work." In A. Scott and M. Storper, eds., *Production, Work, Territory: The Geographical Anatomy of Industrial Capitalism*, pp. 149–71. Boston: Allen and Unwin.

Netting, Robert McC. 1968. *Hill Farmers of Nigeria: Cultural Ecology of the Kofyar of the Jos Plateau*. Seattle: University of Washington Press.

Nietschmann, Bernard. 1987. "The Third World War." *Cultural Survival Quarterly* 11:1–16.

Nollman, Jim. 1990. *Spiritual Ecology: A Guide for Reconnecting with Nature*. New York: Bantam.

Norton, Bryan. 1987. *Why Preserve Natural Variety?* Princeton, N.J.: Princeton University Press.

Novak, David. 1990. "Wealth and Virtue: The Development of Christian Economic Teaching." In Peter Berger, ed., *The Capitalist Spirit: Toward a Religious Ethic of Wealth Creation*, pp. 51–80. San Francisco: ICS.

O'Connor, James. 1987. *The Meaning of Crisis: A Theoretical Introduction*. Oxford: Basil Blackwell.

———. 1989a. "Political Economy and Ecology of Socialism and Capitalism." *Capitalism, Nature, Socialism* 3:33–57.

———. 1989b. "Prospectus." *Capitalism, Nature, Socialism* 3:1–14.

O'Connor, Martin. 1989. "Codependency and Indeterminacy: A Critique of the Theory of Production." *Capitalism, Nature, Socialism* 3:33–57.

Ogden, Joan, and Williams, Robert. 1989. *Solar Hydrogen: Moving beyond Fossil Fuels*. Washington, D.C.: World Resources Institute.

Olson, Mancur. 1973. "Introduction." In M. Olson and H. Landsberg, eds., *The No Growth Society*, pp. 1–14. New York, W. W. Norton.

Olson, Mancur, and Landsberg, Hans, eds. 1973. *The No Growth Society*. New York: W. W. Norton.

Omenn, Gilbert, and Hollaender, Alexander, eds. 1984. *Genetic Control of Environmental Pollutants*. New York: Plenum.

Oppenheimer, Michael, and Boyle, Robert. 1990. *Dead Heat: The Race against the Greenhouse Effect*. New York: Basic Books.

Orians, Gordon. 1990. "Ecological Concepts of Sustainability." *Environment* 32 (9): 10ff.

O'Riordan, Timothy. 1988. "The Politics of Sustainability." In R. K. Turner, ed., *Sustainable Environmental Management: Principles and Practice*, pp. 29–50. London: Bellhaven.

———. 1989. "The Challenge for Environmentalism." In R. Peet and N. Thrift, eds., *New Models in Geography: The Political-Economy Perspective*, pp. 77–104. London: Unwin Hyman.

Ornstein, Robert, and Ehrlich, Paul. 1989. *New World, New Mind: Moving toward Conscious Evolution*. New York: Simon and Schuster.

Paehlke, Robert. 1989. *Environmentalism and the Future of Progressive Politics*. New Haven, Conn.: Yale University Press.

Parnell, Susan. 1991. "Sanitation, Segregation and the Natives (Urban Areas) Act: African Exclusion from Johannesburg's Malay Location, 1897–1925." *Journal of Historical Geography* 17:271–88.

Patterson, James. 1989. "Industrial Waste Reduction." *Environmental Science Technology* 3:1032–38.

Pearce, David, Barbier, Edward, and Markandya, Anil. 1990. *Sustainable Development: Economics and Environment in the Third World*. London: Earthscan.

Pearson, Charles. 1985. *Down to Business: Multinational Corporations, the Environment, and Development*. Washington, D.C.: World Resources Institute.

Peet, Richard. 1989. "World Capitalism and the Destruction of Regional Cultures." In R. J. Johnston and P. J. Taylor, eds., *A World in Crisis? Geographical Perspectives*, pp. 175–99. Oxford: Basil Blackwell.

———. 1991. "The End of History? . . . Or Its Beginning?" *The Professional Geographer* 43:512–19.

Pell, Eve. 1990. "Buying In: How Corporations Keep an Eye on Environmental Groups that Oppose Them—By Giving Them Money." *Mother Jones*, April–May 1990, pp. 23–27.

Pepper, David. 1989. *The Roots of Modern Environmentalism*. London: Routledge.

Perdue, Peter. 1987. *Exhausting the Earth: State and Peasant in Hunan 1500–1850.* Cambridge, Mass.: Council on East Asian Studies, Harvard University.

Perlin, John. 1989. *A Forest Journey: The Role of Wood in the Development of Civilization.* New York: W. W. Norton.

Piasecki, Bruce, and Asmus, Peter. 1990. *In Search of Environmental Excellence: Moving beyond Blame.* New York: Simon and Schuster.

Pickett, S. T. A., and White, P. S., eds. 1985. *The Ecology of Natural Disturbance and Patch Dynamics.* New York: Academic.

Pimenta, João Carlos. 1987. "Multinational Corporations and Industrial Pollution Control in São Paulo, Brazil." In C. Pearson, ed., *Multinational Corporations, Environment, and the Third World,* pp. 198–220. Durham, N.C.: Duke University Press.

Plant, Judith, ed. 1989. *Healing the Wounds: The Promise of Ecofeminism.* Philadelphia: New Society.

Plumwood, Val. 1988. "Women, Humanity and Nature." *Radical Philosophy* 48:6–24.

Polanyi, Karl. 1957. "The Economy as an Instituted Process." In K. Polanyi, C. Arensburg, H. Pearson, eds., *Trade and Market in Early Empires,* pp. 243–70. Glencoe, Ill.: Free Press of Glencoe.

Pollan, Michael. 1991. *Second Nature: A Gardener's Education.* New York: Atlantic Monthly.

Porritt, Jonathon. 1985. *Seeing Green: The Politics of Ecology Explained.* Oxford: Basil Blackwell.

Porter, Gareth, and Brown, Janet Walsh. 1991. *Global Environmental Politics.* Boulder, Colo.: Westview.

Porter, Michael. 1990. *The Competitive Advantage of Nations.* New York: Free Press.

Postrel, Virginia. 1990. "Forget Left and Right, the Politics of the Future will be Growth vs. Green." *Utne Reader,* July–August 1990, pp. 57–58. Excerpted from the Outlook section of the *Washington Post,* April 1, 1990.

Pounds, Norman. 1989. *Hearth and Home: A History of Material Culture.* Bloomington: Indiana University Press.

Pyne, Stephen. 1990. "Firestick History." *Journal of American History* 76:1132–41.

Rambo, A. Terry. 1984. "No Free Lunch: A Reexamination of the Energetic Efficiency of Swidden Agriculture." In A. T. Rambo and P. E. Sajise, eds., *An Introduction to Human Ecology Research on Agricultural Systems in Southeast Asia,* pp. 154–63. Los Banos, Philippines: University of the Philippines at Los Baños, and Honolulu: The East–West Environment and Policy Institute.

———. 1985. *Primitive Polluters: Semang Impact on the Malaysian Tropical Rainforest Ecosystem.* Ann Arbor: Anthropological Papers, Museum of Anthropology, University of Michigan, no. 76.

Rappaport, Roy. 1967. *Pigs for the Ancestors.* New Haven, Conn.: Yale University Press.

———. 1971. "The Flow of Energy in an Agricultural System." *Scientific American* 224:104–15.

———. 1979. *Ecology, Meaning, and Religion.* Berkeley, Calif.: North Atlantic.

Ray, Dixy Lee. 1990. *Trashing the Planet: How Science Can Help Us Deal with Acid Rain, Depletion of Ozone, and Nuclear Waste (among Other Things).* Washington, D.C.: Regenery Gateway.

Redclift, Michael. 1988. "Economic Models and Environmental Values: A Discourse on Theory." In R. K. Turner, ed., *Sustainable Environmental Management: Principles and Practice*, pp. 51–66. London: Bellhaven.

Reddy, Amulya, and Goldemberg, José. 1990. "Energy for the Developing World." *Scientific American* 263:110–18.

Reddy, William. 1984. *The Rise of Market Culture: The Textile Trade and French Society, 1750–1900*. Cambridge: Cambridge University Press.

Reich, Robert. 1987. *Tales of a New America*. New York: Times Books.

———. 1991. *The Work of Nations: Preparing Ourselves for 21st-Century Capitalism*. New York: Alfred A. Knopf.

Reid, Anthony. 1987. "Low Population Growth and Its Causes in Pre-Colonial Southeast Asia." In N. Owen, ed., *Death and Disease in Southeast Asia: Explorations in Social, Medical, and Demographic History*, pp. 33–47. Oxford: Oxford University Press.

———. 1988. *Southeast Asia in the Age of Commerce: Volume One: The Lands below the Winds*. New Haven, Conn.: Yale University Press.

Renfrew, Colin. 1987. *Archeology and Language: The Puzzle of Indo-European Origins*. New York: Cambridge University Press.

Reno, David. n.d. "Indonesia's Transmigration Policy in its Cultural Context." Unpublished research paper, Department of Geography, George Washington University, Washington, D.C..

Repetto, Robert 1985. "Population, Resource Pressure, and Poverty." In Repetto, ed., *The Global Possible: Resources, Development and the New Century*, pp. 131–70. New Haven, Conn.: Yale University Press.

———. 1986. *World Enough and Time: Successful Strategies for Resource Management*. New Haven, Conn.: Yale University Press.

———. 1988. "Subsidized Timber Sales from National Forest Lands in the United States." In R. Repetto and M. Gillis, eds., 1988. *Public Policies and the Misuse of Forest Resources*, pp. 353–84. Cambridge: Cambridge University Press.

———. 1990. *Promoting Environmentally Sound Economic Progress: What the North Can Do*. Washington, D.C.: World Resources Institute.

Repetto, Robert, and Gillis, Malcolm, eds. 1988. *Public Policies and the Misuse of Forest Resources*. Cambridge: Cambridge University Press.

Richards, Paul. 1983. "Ecological Change and the Politics of African Land Use." *African Studies Review* 26:1–71.

Riddell, R. 1981. *Ecodevelopment*. New York: St. Martin's.

Rifkin, Jeremy. 1983. *Algeny*. New York: Viking.

———. 1989. *Entropy: Into the Greenhouse World*. New York: Bantam.

Ringrose, David. 1989. "Towns, Transport and Crown: Geography and the Decline of Spain." In E. Genovese and L. Hochberg, eds., *Geographic Perspectives in History*, pp. 57–80. Oxford: Basil Blackwell.

Rolston, Holmes.1989. *Philosophy Gone Wild*. Buffalo, N.Y.: Prometheus.

Rose, Gillian. 1991. Review of E. Soja, *Postmodern Geographies*, and D. Harvey, *The Conditions of Postmodernity*. *Journal of Historical Geography* 17:118–21.

Rosecrance, Richard. 1990. *America's Economic Resurgence: A Bold New Strategy*. New York: Harper and Row.

Rosenberg, Nathan, and Birdzell, L. E. 1986. *How the West Grew Rich: The Economic Transformation of the Industrial World*. New York: Basic Books.

Ross, Marc, and Steinmeyer, Daniel. 1990. "Energy for Industry." *Scientific American* 263:88–98.

Rostlund, E. 1960. "The Geographic Range of the Historic Bison in the Southeast." *Annals of the Association of American Geographers* 50:395–407.

Roszak, Theodore. 1979. *Person/Planet: The Creative Disintegration of Industrial Society*. Garden City, N.Y.: Anchor.

Routley, Val. 1981. "On Karl Marx as an Environmental Hero." *Environmental Ethics* 3:237–44.

Rubin, Charles. 1989. "Environmental Policy and Environmental Thought: Ruckelshaus and Commoner." *Environmental Ethics* 11:27–51.

Sack, Robert. 1986. *Human Territoriality: Its Theory and History*. Cambridge: Cambridge University Press.

Sagan, Eli. 1974. *Cannibalism*. New York: Harper and Row.

Sagoff, Mark. 1991. "On Making Nature Safe for Biotechnology." In L. Ginzburg, ed., *Assessing Ecological Risks of Biotechnology*, pp. 341–65. Boston: Butterworth–Heinemann.

Sahlins, Marshall. 1972. *Stone Age Economics*. Chicago: Aldine.

———. 1976. *Culture and Practical Reason*. Chicago: University of Chicago Press.

Said, Edward. 1978. *Orientalism*. New York: Vintage.

Sale, Kirkpatrick. 1985. *Dwellers in the Land: The Bioregional Vision*. San Francisco: Sierra Club.

Salleh, Ariel. 1983. "Deeper than Deep Ecology: The Eco-Feminist Connection." *Environmental Ethics* 6:339–45.

Sanday, Peggy. 1986. *Divine Hunger: Cannibalism as a Cultural System*. Cambridge: Cambridge University Press.

Sauer, Carl. 1938 (1963). "Theme of Plant and Animal Destruction in Economic History." In J. Leighly, ed., *Land and Life: A Selection from the Writings of Carl Ortwin Sauer*, pp. 145–54. Berkeley: University of California Press.

Schama, Simon. 1988. *The Embarrassment of Riches*. Berkeley: University of California Press.

Scheffer, Victor. 1991. *The Shaping of Environmentalism in America*. Seattle: University of Washington Press.

Scherr, S. Jacob. 1987. "Hazardous Exports: U.S. and International Policy Developments." In C. Pearson, ed., *Multinational Corporations, Environment, and the Third World*, pp. 129–48. Durham, N.C.: Duke University Press.

Schirmer, Daniel, and Shalom, Stephen, eds. 1987. *The Philippines Reader: A History of Colonialism, Neocolonialism, Dictatorship, and Resistance*. Quezon City, Philippines: KEN.

Schive, Chi. 1990. "The Next Stage of Industrialization in Taiwan and Korea." In G. Gereffi and D. Wyman, eds., *Manufacturing Miracles: Paths of Industrialization in Latin America and East Asia*, pp. 267–91. Princeton, N.J.: Princeton University Press.

Schoenberger, Erica. 1990. "U.S. Manufacturing Investments in Western Europe: Markets, Corporate Strategy, and Competitive Environment." *Annals of the Association of American Geographers* 80:379–94.

Schumacher, E. F. 1973. *Small Is Beautiful*. New York: Harper and Row.

Schumpeter, Joseph. 1942. *Capitalism, Socialism, and Democracy*. New York: Harper Brothers.

Scott, Allen, and Storper, Michael, eds. 1986. *Production, Work, Territory: The Geographical Anatomy of Industrial Capitalism*. Boston: Allen and Unwin.

Scott, Geoffrey. 1977. "The Role of Fire in the Creation and Maintenance of Savanna in the Montana of Peru." *Journal of Biogeography* 4:143–67.

Seavoy, Ronald. 1975. "The Origins of Tropical Grasslands in Kalimantan, Indonesia." *Journal of Tropical Geography* 40:48–52.

Seldon, Arthur. 1990. *Capitalism*. Oxford: Basil Blackwell.

Semple, Ellen Churchill. 1911. *Influences of Geographic Environment*. New York: H. Holt.

Shiva, Vandana. 1989. "Women, Ecology and Development." In J. Plant, ed., *Healing the Wounds: The Promise of Ecofeminism*, pp. 80–90. Philadelphia: New Society.

Silver, Timothy. 1990. *A New Face on the Countryside: Indians, Colonists, and Slaves in South Atlantic Forests, 1500–1800*. Cambridge: Cambridge University Press.

Simmons, I. G. 1989. *Changing the Face of the Earth: Culture, Environment, History*. Oxford: Basil Blackwell.

Simon, Julian. 1987. *The Ultimate Resource*. Princeton, N.J.: Princeton University Press.

Simon, Julian, and Kahn, Herman, eds. 1984. *The Resourceful Earth: A Response to Global 2000*. New York: Basil Blackwell.

Singer, Peter. 1975. *Animal Liberation: A New Ethic for Our Treatment of Animals*. New York: Avon.

Sisk, John. 1991. "Notes from the Ecological Depths." *The Georgia Review* 45 (2):235–45.

Smil, Vaclav. 1984. *The Bad Earth: Environmental Degradation in China*. Armonk, N.Y.: M. E. Sharpe.

Smith, Gar (compiler). 1990. "50 Difficult Things You Can Do to Save the Earth." *Utne Reader* July–August 1990, p. 56.

Smith, J. Russell. 1953. *Tree Crops: A Permanent Agriculture*. New York: Devin-Adair.

Soulé, Michael. 1980. "Conservation Biology: Its Scope and Challenge." In M. Soulé and B. Wilcox, eds., *Conservation Biology: An Evolutionary-Ecological Perspective*. Sunderland, Mass.: Sinauer.

Stallings, Barbara. 1990. "The Role of Foreign Capital in Economic Development." In G. Gereffi and D. Wyman, eds., *Manufacturing Miracles: Paths of Industrialization in Latin America and East Asia*, pp. 55–89. Princeton, N.J.: Princeton University Press.

Starhawk. 1989. "Feminist Earth-Based Spirituality and Ecofeminism." In J. Plant, ed., *Healing the Wounds: The Promise of Ecofeminism*, pp. 174–88. Philadelphia: New Society.

Steward, J. 1955. *Theory of Culture Change*. Urbana: University of Illinois Press.

Stewart, Omer. 1956. "Fire as the First Great Force Employed by Man." In W. L. Thomas, ed., *Man's Role in Changing the Face of the Earth: Volume One*, pp. 115–33. Chicago: University of Chicago Press.

Stone, Christopher. 1988. "Moral Pluralism and the Course of Environmental Ethics." *Environmental Ethics* 10:139–54.

Strathern, Marilyn, ed. 1987. *Dealing with Inequality: Analyzing Gender Relations in Melanesia and Beyond*. Cambridge: Cambridge University Press.

Strathern, Marilyn. 1991. "Introduction." In M. Godelier and M. Strathern, eds., *Big*

Men and Great Men: Personifications of Power in Melanesia, pp. 1–4. Cambridge: Cambridge University Press.

Strauss, William, and Howe, Neil. 1991. *Generations: The History of America's Future, 1584–2069*. New York: William Morrow.

Tarrant, John. 1990. "Food Policy Conflicts in Food-Deficit Countries." *Progress in Human Geography* 14:467–87.

Taylor, Paul. 1986. *Respect for Nature: A Theory of Environmental Ethics*. Princeton, N.J.: Princeton University Press.

Taylor, Peter. 1989. "The World Systems Project." In R. J. Johnston and P. J. Taylor, eds., *A World in Crisis? Geographical Perspectives*, pp. 333–54. Oxford: Basil Blackwell.

Terborgh, John. 1989. *Where Have all the Birds Gone?: Essays on the Biology and Conservation of Birds that Migrate to the American Tropics*. Princeton, N.J.: Princeton University Press.

Thomas, William, ed. 1956. *Man's Role in Changing the Face of the Earth*. Chicago: University of Chicago Press.

Thrift, Nigel. 1989. "The Geography of International Economic Disorder." In R. J. Johnston and P. J. Taylor, eds., *A World in Crisis? Geographical Perspectives*, pp. 16–78. Oxford: Basil Blackwell.

Thrupp, Lori Ann. 1989. "Discussion. The Struggle for Nature: Replies (2)." *Capitalism, Nature, Socialism* 3:169–74.

Thurow, Lester. 1985. *The Zero Sum Solution: Building a World-Class American Economy*. New York: Simon and Schuster.

Tienda, Marta. 1984. "Community Characteristics, Women's Education, and Fertility in Peru." *Studies in Family Planning* 15:162–69.

Tierney, John. 1990. "Betting the Planet." *New York Times Magazine*, December 2, 1991, pp. 52ff.

Tokar, Brian. 1987. *The Green Alternative: Creating an Ecological Future*. San Pedro, Calif.: R. and E. Miles.

Tolman, Charles. 1981. "Karl Marx, Alienation, and the Mastery of Nature." *Environmental Ethics* 3:63–74.

Torgovnick, Marianna. 1990. *Gone Primitive: Savage Intellects, Modern Lives*. Chicago: University of Chicago Press.

Tough, Frank. 1990. "Indian Economic Behavior, Exchange and Profits in Northern Manitoba during the Decline of Monopoly, 1870–1930." *Journal of Historical Geography* 16:385–401.

Toulmin, Stephen. 1990. *Cosmopolis: The Hidden Agenda of Modernity*. New York: Free Press.

Tuan, Yi-Fu. 1974. "Discrepancies between Environmental Attitudes and Behavior." In Spring and D. Spring, eds., *Ecology and Religion in History*, New York: Harper and Row.

Tucker, William. 1982. *Progress and Privilege: America in the Age of Environmentalism*. Garden City, N.Y.: Anchor.

Turnbull, Colin. 1972. *The Mountain People*. New York: Simon and Schuster.

Turner, R. Kerry. 1988. "Sustainability, Resource Conservation and Pollution Control: An Overview." In R. K. Turner, ed., *Sustainable Environmental Management: Principles and Practice*, pp. 1–28. London: Bellhaven.

Vance, James. 1970. *The Merchant's World: The Geography of Wholesaling*. Englewood Cliffs, N.J.: Prentice-Hall.

———. 1977. *This Scene of Man: The Role and Structure of the City in the Geography of Western Civilization*. New York: Harper's College Press.

van den Berghe, P. 1981. *The Ethnic Phenomenon*. New York: Elsevier.

van Wolferen, Karel. 1989. *The Enigma of Japanese Power: People and Politics in a Stateless Nation*. London: Macmillan.

Walker, Richard. 1988. "The Geographical Organization of Production Systems." *Environment and Planning D: Society and Space* 6:377–408.

———. 1989. "What's Left to Do?" *Antipode* 21:133–65.

Wallerstein, I. 1974. *The Modern World System: Capitalist Agriculture and the Origins of the European World Economy in the Sixteenth Century*. New York: Academic.

Warren, James Francis. 1985. *The Sulu Zone 1768–1898: The Dynamics of External Trade, Slavery, and Ethnicity in the Transformation of a Southeast Asian Maritime State*. Quezon City, Philippines: New Day.

Warren, Karen. 1987. "Feminism and Ecology: Making Connections." *Environmental Ethics* 9:3–20.

———. 1990. "The Power and Promise of Ecological Feminism." *Environmental Ethics* 12:125–47

Washbrook, David. 1990. "South Asia, the World System, and Capitalism." *Journal of Asian Studies* 49:479–508.

Wattenberg, Ben. 1984. *The Good News Is the Bad News Is Wrong*. New York: Simon and Schuster.

———. 1987. *The Birth Dearth*. New York: Pharos.

Watts, Michael. 1988. "Deconstructing Determinism: Marxisms, Development Theory, and a Comradely Critique of *Capitalist World Development*." *Antipode* 20: 142–63.

Weinberg, Carl, and Williams, Robert. 1990. "Energy from the Sun." *Scientific American* 263:146–55.

Wheeler, James, and Mitchelson, Ronald. 1989. "Information Flows among Major Metropolitan Areas in the United States." *Annals of the Association of American Geographers* 79:523–43.

White, Lynn. 1967 (1973). "The Historical Roots of our Ecological Crisis." In I. Barbour, ed., *Western Man and Environmental Ethics*, pp. 18–30. Reading, Mass.: Addison-Wesley.

Whitehand, J. W. R. 1990. "An Assessment of 'Progress.'" *Progress in Human Geography* 14:12–23.

Whittlesey, Derwent. 1954. "The Regional Concept and the Regional Method." In P. James and C. Jones, eds., *American Geography: Inventory and Prospective*, pp. 10–69. Syracuse, N.Y.: Syracuse University Press.

Wilken, Gene. 1987. *Good Farmers: Traditional Agricultural Resource Management in Mexico and Central America*. Berkeley: University of California Press.

Wilkinson, Richard. 1988. "The English Industrial Revolution." In D. Worster, ed., *The Ends of the Earth: Perspectives on Modern Environmental History*, pp. 80–102. Cambridge: Cambridge University Press.

Williams, Raymond. 1983. *Culture and Society 1780–1950*. New York: Columbia University Press.

Willis, K. G., and Benson, J. F. 1988. "Valuation of Wildlife: A Case Study of the Upper Teesdale Site of Specific Scientific Interest and Comparison of Methods in Environmental Economics." In R. K. Turner, ed., *Sustainable Environmental Management: Principles and Practice*, pp. 243–64. London: Bellhaven.

Wilmsen, Edwin. 1983. "The Ecology of Illusion: Anthropological Foraging in the Kalahari." *Reviews in Anthropology* 10:9–20.

Wilmsen, Edwin, and Denbow, James. 1990. "Paradigmatic History of San-Speaking Peoples and Current Attempts at Revision." *Current Anthropology* 31:489–524.

Winner, Langdon. 1986. *The Whale and the Reactor: A Search for Limits in Age of High Technology*. Chicago: University of Chicago Press.

Wirsing, Rolf. 1985. "The Health of Traditional Societies and the Effects of Acculturation." *Current Anthropology* 26:303–22.

Wolpe, H. 1972. "Capitalism and Cheap Labour Power: From Segregation to Apartheid." *Economy and Society* 6.

Wood, Harold. 1985. "Modern Pantheism as an Approach to Environmental Ethics." *Environmental Ethics* 7:151–64.

Wood, Joseph. 1991. " 'Build, Therefore, Your Own World': The New England Village as Settlement Ideal." *Annals of the Association of American Geographers* 81:32–50.

World Bank. 1990. *The World Bank Atlas 1990*. Washington, D.C.: World Bank.

Worster, Donald. 1985. *Rivers of Empire: Water, Aridity, and the Growth of the American West*. New York: Pantheon.

———. 1988. "Appendix: Doing Environmental History." In D. Worster, ed., *The Ends of the Earth: Perspectives on Modern Environmental History*, pp. 289–307. Cambridge: Cambridge University Press.

———. 1990. "Transformations of the Earth: Toward an Agro-ecological Perspective in History." *Journal of American History* 76:1087–1106.

Wright, Erik. O. 1983. "Gidden's Critique of Marx." *New Left Review* 138:11–35.

Young, John. 1990. *Sustaining the Earth*. Cambridge, Mass.: Harvard University Press.

Yun-Shik, Chang. 1991. "The Personalist Ethic and the Market in Korea." *Comparative Studies in Society and History* 33:106–29.

Zimmerer, Karl. Forthcoming. "Ecology." In C. Earle and M. Kenzer, eds., *Conceptual Thinking in Human Geography*. London: Routledge.

Index

Martin Lewis is Assistant Professor in the Department
of Geography and Regional Science, George Washington
University. He is the author of *Wagering the Land:
Ritual, Capital, and Environmental Degradation in the
Cordillera of Northern Luzon 1900–1986.*

Library of Congress Cataloging-in-Publication Data
Lewis, Martin W.
Green delusions : an environmentalist critique of
radical environmentalism / Martin W. Lewis.
Includes bibliographical references and index.
ISBN 0-8223-1257-3
1. Environmental policy—Citizen participation.
2. Radicalism.   I. Title.
HC79.E5L48   1992
363.7'057—dc20   92-5671
CIP